T0320405

Explainable Artificial Intelligence (XAI) in Healthcare

This book highlights the use of explainable artificial intelligence (XAI) for healthcare problems, in order to improve trustworthiness, performance, and sustainability levels in the context of applications.

Explainable Artificial Intelligence (XAI) in Healthcare adopts the understanding that AI solutions should not only have high accuracy performance, but also be transparent, understandable, and reliable from the end user's perspective. The book discusses the techniques, frameworks, and tools to effectively implement XAI methodologies in critical problems of the healthcare field. The authors offer different types of solutions, evaluation methods, and metrics for XAI and reveal how the concept of explainability finds a response in target problem coverage. The authors examine the use of XAI in disease diagnosis, medical imaging, health tourism, precision medicine, and even drug discovery. They also point out the importance of user perspectives and the value of the data used in target problems. Finally, the authors also ensure a well-defined future perspective for advancing XAI in terms of healthcare.

This book will offer great benefits to students at the undergraduate and graduate levels and researchers. The book will also be useful for industry professionals and clinicians who perform critical decision-making tasks.

Biomedical and Robotics Healthcare

Series Editors: Utku Kose, Jude Hemanth, and Omer Deperlioglu

For more information about this series, please visit: www.routledge.com/
Biomedical-and-Robotics-Healthcare/book-series/BRHC

Explainable Artificial Intelligence (XAI) in Healthcare

Edited by
Utku Kose, Nilgun Sengoz, Xi Chen and
Jose Antonio Marmolejo Saucedo

CRC Press
Taylor & Francis Group
Boca Raton London New York

CRC Press is an imprint of the
Taylor & Francis Group, an **informa** business

Designed cover image: © Shutterstock Images

First edition published 2024
by CRC Press
2385 NW Executive Center Drive, Suite 320, Boca Raton FL 33431

and by CRC Press
4 Park Square, Milton Park, Abingdon, Oxon, OX14 4RN

CRC Press is an imprint of Taylor & Francis Group, LLC

© 2024 selection and editorial matter, Utku Kose, Nilgun Sengoz, Xi Chen and Jose Antonio Marmolejo Saucedo; individual chapters, the contributors

Library of Congress Cataloging-in-Publication Data
Names: Kose, Utku, 1985– editor. | Sengoz, Nilgun, editor. |
Chen, Xi (Researcher in bioinformatics), editor. |
Marmolejo Saucedo, Jose Antonio, editor.
Title: Explainable artificial intelligence (XAI) in healthcare /
edited by Utku Kose, Nilgun Sengoz, Xi Chen, and Jose Antonio Marmolejo Saucedo.
Description: First edition. | Boca Raton : CRC Press, 2024. |
Series: Biomedical and robotics healthcare |
Includes bibliographical references and index. |
Identifiers: LCCN 2023051962 (print) | LCCN 2023051963 (ebook) |
ISBN 9781032543703 (hbk) | ISBN 9781032546674 (pbk) | ISBN 9781003426073 (ebk)
Subjects: LCSH: Artificial intelligence–Medical applications.
Classification: LCC R859.7.A78 E97 2024 (print) | LCC R859.7.A78 (ebook) |
DDC 610.285–dc23/eng/20240301
LC record available at https://lccn.loc.gov/2023051962

ISBN: 9781032543703 (hbk)
ISBN: 9781032546674 (pbk)
ISBN: 9781003426073 (ebk)

DOI: 10.1201/9781003426073

Typeset in Times New Roman
by Newgen Publishing UK

Contents

Foreword

Humankind has been always involved in understanding nature and developing solutions for problems encountered. In time, these efforts have resulted in the creation of technology concepts involving knowledge, actions, and tools. Thanks to technological developments, it has been more critical to increase the modernity level of societies and shape the future. Among the developments, there have been some revolutionary transitions from one technological era to another. After the rise of computer and computer technologies in the 20th century, there was a remarkable search for more automated tools for processing information and ensuring adaptive solutions targeting real-world problems. As a result, artificial intelligence technology, which is aimed at designing and developing advanced algorithmic systems, was taken to stage. Today, it is the reason behind all adaptive intelligent systems, which ensure flexible data processing capabilities and even design the working mechanisms of smart devices surrounding us.

It is remarkable that intelligent systems have been effective in terms of achieveing high accuracy prediction, recognition, and data discovery solutions for all fields of modern life. Since the first appearance of the artificial intelligence, healthcare has been intensively in touch with such solutions and many breathtaking advancements have been made so far. Today, we are able to talk about robotic operations, early disease diagnosis, drug discoveries, treatment planning, and precise medicine efforts, which would not be real without technological as well as scientific contributions by artificial intelligence. However, more advancements in technological tools have resulted in more complicated systems. Because of the increased complexies, it is now difficult for humans to interpret and evaluate how such systems are working and how far they are able to work in a sustainable way. This issue is more vital for healthcare applications including intelligent systems. Without enough idea about the inside mechanisms and inferencing flows, any intelligent system, which tends to have errors in data processing steps, may result in harmful problems for the human side. Thus, recent research efforts came up with Explainable Artificial Intelligence (XAI) where additional components are widely used to ensure interpretability or explainability for data processing steps of intelligent systems.

Titled as *Explainable Artificial Intelligence (XAI) in Healthcare*, this book gives a recent view on how XAI research has been widely applied in terms of different domains of healthcare. It is critical that healthcare applications include different type of data (from raw, to image, time series, etc.), and today's intelligent systems are mostly run over hybrid machine learning or advanced deep learning models. So, the general coverage of the book is organized in a way to reflects the role of XAI in complicated problem solutions and black-box systems. It is nice to see that there are comprehersive review works, which ensure a good opportunity for readers to follow up on the up-to-date literature and understand more about what has been done so far for advancing the XAI in healthcare problems. Furthermore, the book covers critical topics such as drug discovery and medical imaging, which are triggering new revolutions in digital healthcare. It is also very important to see that all chapters ensure introductive explanations for their target problems and increase the level of technical details associated with their target research points. I believe that the book will be effectively used by not only researchers but also professionals from both public and private healthcare sectors. Also, undergraduate and graduate students can refer to the book for designing their further research studies. I would like to thank to the respectful editors team, Dr. Kose, Dr. Sengoz, Dr. Chen, and Dr. Marmolejo Saucedo, for their efforts to make this book real. I wish also every reader to enjoy their journey through the pages of this great work. I highly suggest all readers read every chapter carefully to understand how the XAI and healthcare relation is expanding and how trustworthy intelligent systems of the future may be developed.

Dr. Omer Deperlioglu
Prof. Dr. of Computer Technologies,
Afyon Kocatepe University, Turkey

Preface

In today's world, artificial intelligence-based systems are more complicated than the ones even 20 years ago. That's because of the data era, digital transformation, and advancements of real-world problems, which appear as a result of new technologies. It can be understood that there is a loop in terms of new technologies and the new problems requiring newer technologies. In the context of this flow, artificial intelligence has been an important actor to ensure comprehensive intelligent systems for successful outcomes for even difficult problems. However, there has been a tradeoff when it comes to complexity of intelligent systems and improved success rates. As a result, today's intelligent systems, which are mostly run over deep learning models, are difficult to be interpreted or simply black box for the user side. This is a critical issue since there have already been anxieties regarding thrustworthiness of advancing intelligent systems. So, the literature of artificial intelligence has been enrolled in research on using interpretable machine learning techniques inside deep learning or simply introducing new methods to be integrated into deep learning models for ensuring explainable interfaces. These efforts have resulted in a new research area called explainable artificial intelligence (XAI) to taking an intensive place in research studies since especially 2010s.

XAI has received a great deal of interest as it allows development of trustworthy intelligent systems that can explain how they create bridges between inputs and outcomes (outputs). Since trust for intelligent systems is critical for humans, XAI usage has gained a great momentum recently. By targeting the well-being of humans and ensuring healthy future generations, the medical field has been highly connected to XAI-based solutions. Since healthcare actually has a wider scope including all living organisms in nature, contributions by XAI have been very useful in building trustworthy intelligent systems for all kind of problems. In this context, healthcare problems have also been widely connected to XAI use and better human-artificial intelligence relations. It is believed that the gap for human capabilities because of black-box intelligent systems can be effectively closed as a result of XAI integration into the intelligent systems of the present and future.

Explainable Artificial Intelligence (XAI) in Healthcare provides an edited volume in which recent examples of trustworty XAI are reported accordingly. Since healthcare applications may have different types of data analysis and artificial intelligence usage, the chapters included in the book were carefully reviewed and selected to inform readers about the latest advancements. The audience of this book is not only researchers with XAI interest but also experts from the sector as well as MSc, PhD, and PostDoc students working in the context of XAI in healthcare. In order to build a valuable reference work, the book explains the essentials of XAI and report on the latest state of the literature for starting level readers. This volume covers use cases of artificial intelligence in healthcare (Chapter 1); general evaluation of XAI use in biomedical (Chapter 3); XAI use for drug discovery (Chapter 4, Chapter 11, and Chapter 12); the general state of XAI for medical imaging-based research (Chapter 5 and Chapter 6); a review of XAI advancement in the literature (Chapter 7); XAI use cases for disease diagnosis (Chaper 8); digital twin, personalized medicine, and XAI connections (Chapter 9); and XAI usage for medical tourism applications (Chapter 10). As a whole, the book considers critical topics as well as current research in the artificial intelligence-based healthcare domain. Due to the general coverage and up-to-date knowledge, the book will be useful as a reference to research, lectures, and general interest for a long time coming.

As the editors, we would like to thank all the authors for their contributions to organize such a timely work. Additionally, our special thanks are to Prof. Dr. Omer Deperlioglu for his kind review and positive words for this book. We always await any ideas and suggestions for the book and further projects. Thanks!

Utku Kose, Nilgun Sengoz, Xi Chen, and Jose Antonio Marmolejo Saucedo

About the Editors

Utku Kose received a B.S. degree in 2008 from computer education of Gazi University, Turkey as a faculty valedictorian. He received a M.S. degree in 2010 from Afyon Kocatepe University, Turkey in the field of computer and a D.S./Ph. D. degree in 2017 from Selcuk University, Turkey in the field of computer engineering. Currently, he is an Associate Professor at Suleyman Demirel University, Turkey. Kose has given lectures at other higher education institutions such as Gazi University and Istanbul Arel University. He also works as a Visiting Researcher at the University of North Dakota, USA and holds the Honorary Professor of Artificial Intelligence title at ITM (SLS) Baroda University, India. He has more than 300 publications including articles, authored and edited books, proceedings, and reports. He is also on the editorial boards of many scientific journals and serves as one of the editors of the *Biomedical and Robotics Healthcare* (CRC Press) and *Computational Modeling Applications for Existential Risks* (Elsevier) book series. His research interests include artificial intelligence, machine ethics, artificial intelligence safety, biomedical applications, optimization, the chaos theory, distance education, e-learning, computer education, and computer science.

Nilgun Sengoz received her Bsc degree in industrial engineering from Atılım University, Ankara, Turkey in 2008 and an Msc degree in industrial engineering from Suleyman Demirel University, Isparta, Turkey in 2016. She received her PhD at the Department of Computer Engineering, Isparta, Turkey in 2022. She is currently an Assistant Professor at Burdur Mehmet Akif Ersoy University, Turkey. Her areas of interest are artificial intelligence, machine learning and deep learning, medical image processing, and also human computer interaction.

Xi Chen received his MA degree in Statistics and PhD degree in Biometrics from University of Kentucky, USA. He was awarded Microsoft Azure Research Funding in 2016 and got the Google Challenge Scholarship in 2018. He is a SAS certified advanced programmer and NVidia certified Deep Learning/Cuda instructor. Dr. Chen's research interests include Deep Learning, XAI, and medical applications including disease diagnosis, and genomics.

Jose Antonio Marmolejo Saucedo is a Professor at National Autonomous University of Mexico, Mexico. His research is on operations research, largescale optimization techniques, computational techniques, analytical methods for planning, operations, and control of electric energy and logistic systems, sustainable supply chain design, and digital twins in supply chains. He is a member of the Network for Decision Support and Intelligent Optimization of Complex and Large Scale Systems, Mexican Society for Operations Research and System Dynamics Society. He is the author of more than thirty research articles in science citation index journals, books, book chapters, conference proceedings, and presentations.

Contributors

Enes Açıkgözoğlu
Isparta University of Applied Sciences
Isparta, Turkey

Tuba Aftab
Dept. of Biosciences
COMSATS University Islamabad (CUI)
Park Road, Islamabad, Pakistan

Haroon Ahmed
Dept. of Biosciences
COMSATS University Islamabad (CUI)
Park Road, Islamabad, Pakistan

Shalom Akhai
Chandigarh College of Engineering
CGC Jhanjeri, Mohali, India 140307

Bekir Aksoy
Isparta University of Applied Sciences
Isparta, Turkey

Sema Çayır
Isparta University of Applied Sciences
Isparta, Turkey

Xi Chen
Meta
Burlingame, CA, USA

Omer Deperlioglu
Dept. of Computer Technologies
Afyon Kocatepe University
Afyonkarahisar, Turkey

Mevlüt Ersoy
Dept. of Computer Engineering
Suleyman Demirel University
Isparta, Turkey

Remzi Gürfidan
Isparta University of Applied Sciences
Isparta, Turkey

Mansoor Hussain
Dept. of Biosciences
COMSATS University Islamabad (CUI)
Park Road, Islamabad, Pakistan

Waseem Ullah Jan
Dept. of Biosciences
COMSATS University Islamabad (CUI)
Park Road, Islamabad, Pakistan

Kevser Kubra Kirboga
Dept. of Bioengineering
Bilecik Seyh Edebali University
Bilecik, Turkey

Gamze Kose
Dept. of Health Tourism
Aydin Adnan Menderes University
Aydin, Turkey

Utku Kose
Dept. of Computer Engineering
Suleyman Demirel University
Isparta, Turkey
College of Engineering & Mines
University of North Dakota
Grand Forks, ND, USA

Satish Kumar
Baba Ghulam Shah Badshah University
India

Faisal Rasheed Lone
SCSE
VIT Bhopal University
Madhya Pradesh, India

Roohie Naaz Mir
Dept. of Computer Science &
  Engineering
National Institute of Technology
  Srinagar
Jammu & Kashmir, India

Azra Nazir
SCSE
VIT Bhopal University
Madhya Pradesh, India

Shaima Qureshi
Dept. of Computer Science &
  Engineering
National Institute of Technology
  Srinagar
Jammu & Kashmir, India

M. Ahsan Saeed
Dept. of Biosciences
COMSATS University Islamabad (CUI)
Park Road, Islamabad, Pakistan

Ferdi Sarac
Dept. of Computer Engineering
Suleyman Demirel University
Isparta, Turkey

Jose Antonio Marmolejo Saucedo
National Autonomous University
  of Mexico
Mexico

Nilgun Sengoz
Burdur Mehmet Akif Ersoy University
Burdur, Turkey

Naseer Ali Shah
Dept. of Biosciences
COMSATS University Islamabad (CUI)
Park Road, Islamabad, Pakistan

K. Aditya Shastry
Nitte Meenakshi Institute of Technology
Bengaluru, India

İlhan Uysal
Burdur Mehmet Akif Ersoy University
Burdur, Turkey

Asma Yousaf
Dept. of Biosciences
COMSATS University Islamabad (CUI)
Park Road, Islamabad, Pakistan

1 Artificial Intelligence for Healthcare Applications
A Review

K. Aditya Shastry

1.1 INTRODUCTION

AI is a technological field that employs technological devices to study and create the concept, technique, methodology, and software application for simulating, extending, and expanding cognitive abilities. Alan Turing pioneered the idea of AI in 1950; he established the Turing test and characterised AI as comparable to but more complicated than the human mind (Liu et al., 2021).

There has been a rise in the use of AI in recent times, particularly since the advent of DL (a set of training methods that form the backbone of the latest era of AI skill, with the capability to efficiently gain knowledge from big-data-analysis and afterwards artificially and autonomously decide based on that expertise) (Mintz & Brodie, 2019). This type of AI uses a wide variety of neural networks. Now that AI is being used in so many different areas and contributing so much to technical advancements, a new notion has emerged: AI plus. The term "AI plus" refers to the practise of integrating AI's technological advances and accomplishments with more conventional business models to boost productivity, innovation, and growth (Kaul et al., 2020). Researchers in the field of AI have found that the ratio of benefits to resources invested in medical is higher than in any other sector studied so far. AI combined with healthcare promotes radical change to the established healthcare paradigm. Medical applications of AI have garnered a lot of interest, too, because of their promising future (Patel et al., 2009).

Rapid innovations in AI technology in healthcare have prompted the question of whether AI technologies may eventually replace human physicians. Realistically, AI technologies might not be capable of replacing human physicians, but they could assist them in achieving improved outcomes and superior precision in the health industry. The accessibility of health records is a vital aspect in the growth of these AI medicinal applications (Manne & Kantheti, 2021). AI is a set of techniques, not a single technology. A few of these techniques, such as machine learning (ML), are often utilised in medicine. ML is a procedure in which algorithms are trained utilising pre-existing information so that, based on their pre-learning, they can recognise the

DOI: 10.1201/9781003426073-1

test input whenever fed the information used for testing. One of the most frequent forms of AI is ML (Lee et al., 2018)

AI software, comprising computational ML and ability for self, generates new potential for innovation in a variety of fields, particularly banking, medicine, industry, commerce, distribution network, transportation, and energy (Esmaeilzadeh, 2020; López-Robles et al., 2019). AI could be implemented as clinical information systems (CDSs) to assist physician testing and therapy selections and to do inhabitant predictive analytics (Brufau et al., 2019). The advancement of AI-centered products has become a key component of the strategies of several businesses (Coombs et al., 2020). Recent research has been motivated by the significant alterations brought about by AI to analyse the ramifications and repercussions of the technologies and the resultant effects of AI. Nonetheless, this goal requires comprehensive knowledge of the dynamics that influence the acceptability of AI-centered solutions by prospective customers in various service and manufacturing industries. Prior studies emphasise the significance of AI in healthcare, particularly healthcare analytics (Khanna et al., 2020). AI can deliver enhanced patient safety, diagnostics, and analysis of medical information (Dreyer & Allen, 2018). AI knowledge applied for breast cancer diagnosis minimises human errors, according to research (Houssami et al., 2017), but several connected moral and social trust factors, as well as AI dependency, have still to be created. The implementation of AI-motivated suggestions in the medical domain may differ from other industries due to the sensitivity of health information and the susceptibility of customers to medication error.

The Food and Drug Administration (FDA) authorised the first AI equipment to detect diabetic retinopathy in the absence of a human physician in the United States in April 2018 (Laï et al., 2020). A growing number of healthcare service providers are investing in the development of AI incorporated in mobile medical gadgets or health apps in order to enhance patient safety, boost treatment quality, promote healthcare adminstration, and reduce medical expenses. Nevertheless, prior research indicates that not all people are prepared to accept the usage of healthcare AI devices (Laï et al., 2020). Effective execution of AI-based solutions necessitates a thorough analysis of user values and practices regarding AI (Brufau et al., 2019). Therefore, engaging in AI technology before understanding the attitudes and acceptance of intended consumers might result in a loss of assets and/or customers. This is certainly relevant in the healthcare sector, wherein patient involvement is regarded as one of the most important criterion indicators. If individuals do not consider connecting with an intelligent Healthcare gadget as beneficial, they might seek contact with doctors, causing AI-based gadgets to go underutilised. Knowing the choice factors and obstacles that lead to acceptance or rejection of the usage of AI-based gadgets in health system is crucial for health workers and organisations that want to adopt and/or grow the application of AI-based gadgets in health systems (Esmaeilzadeh, 2020).

This chapter examines the relevant use cases of AI in medicine during recent years.

1.2 AI METHODS IN HEALTHCARE

The ever-present AI technologies in corporate industry and daily life are being used progressively in medicine. The application of AI in medicine has the ability to aid health professionals in a variety of patient care and administration operations, enabling them to adapt current methods and meet challenges more rapidly. While AI and healthcare innovations have a substantial impact on the healthcare industry, the strategies they assist in can differ considerably across clinics and other medical organisations. Though a number of publications on AI in healthcare suggest that AI in healthcare could operate as well as or superior to people at operations, including analysis of diseases, it may be many years until AI in healthcare substitutes people for most medical duties (Manne et al., 2021).

Figure 1.1 shows the commonly used AI methods in healthcare.

Prior to visiting a physician, a patient may interact with a machine as part of normal clinical practice in the coming years. With advancements in AI, it seems that the era of wrong diagnosis and addressing clinical manifestations as opposed to their underlying cause may soon be a thing of the past. The accumulating information recorded in hospitals and kept in electronic medical records via common tests and diagnostic imaging enables additional applications of AI and information-driven medicine with superior performance. These technologies have altered and will continue to alter physicians' and academics' diagnosis strategies (Basu et al., 2020).

Although some methods could operate on par with or even better than physicians in a number of activities, they have not yet been fully incorporated into normal clinical practise. This is due to several legal hurdles that must be resolved initially, even if these programs have the potential to significantly influence healthcare and increase

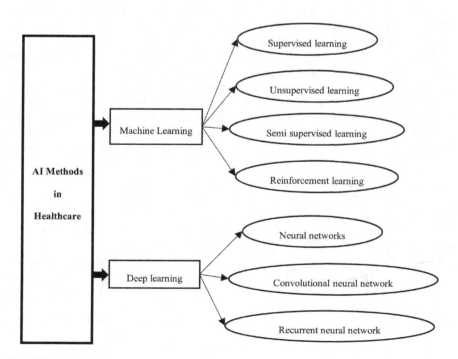

FIGURE 1.1 AI Methods in Healthcare.

the effectiveness of surgical treatments. AI methods must learn how to perform their tasks just like physicians do by spending years in medical college, completing tasks and clinical examinations, obtaining marks, and learning from errors. In principle, jobs that necessitate human intelligence to accomplish, such as pattern and voice recognition, image processing, and decision-making processes, could be performed by AI methods. Nevertheless, when providing a visual to an algorithm, for instance, people must directly instruct the machine about what to look for in the visual. In summary, AI algorithms are excellent at completing laborious jobs and occasionally exceed individuals at the activities they have been programmed to execute (Amisha et al., 2019).

In order to build a successful AI system, computers are often given information, implying that every piece of data has a label or annotation that the method can recognise (Figure 1.2). After the method has been subjected to enough different sets of pieces of information and associated descriptions, its performance is assessed to confirm its correctness, just as students are administered tests. These algorithm "exams" typically entail the input of testing data for which the developers previously have the answers, enabling researchers to evaluate the methods' ability to identify the proper response. Depending on the results of assessment, the algorithms might be updated, supplied additional information, or implemented to assist the program's author in making judgments (Briganti et al., 2020).

Figure 1.2 depicts an AI approach that could help to analyze anatomy of a hand. The intake is a collection of hand x-rays, and the result is a map indicating where omitted hand components ought to be. In this instance, the prototype is the arm shape which may be extended to other imagery. This might help doctors to determine where to repair a limb or insert a replacement (Gomez et al., 2019).

There are numerous methods capable of data-driven learning. Input data for the majority of AI applications in medicine is quantitative (like pulse rate or hypertension) or graphics based (such as MRI scans or imaging of tumour clinical specimens). After learning from the information, the methods allow to determine whether it's a likelihood. For instance, the usable outcome might be the chance of developing a vascular thrombosis based upon pulse rate as well as hypertension information, or the malignant or benign classification of an imaging tissue sample. In clinical

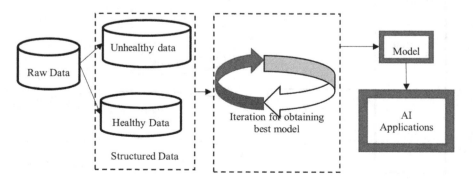

FIGURE 1.2   AI Algorithms.

uses, the effectiveness of an approach on a diagnostic activity is compared to that of a doctor to establish the method's capability and utility in the hospital (Secinaro et al., 2021).

Several clinical problems are ideal for AI applications due to advances in computational capabilities and the massive quantities of information produced by healthcare systems. Two recent implementations of precise and medically useful techniques that make detection easier both for patients and physicians are described below.

This first method is one of numerous extant instances of a computer which surpasses clinicians in imagery categorisation tasks. An AI programme known as DLAD (Deep Learning-based Automatic Detection) to analyse chest images and identify aberrant neurogenesis, like possible malignancies, was created. In a comparison of the algorithm's efficiency against that of several doctors' detecting abilities on the identical images, the programme surpassed 17 of 18 physicians (Meskó & Görög, 2020).

Scientists at Google AI Healthcare established Lymph Node Assistant (LYNA) to detect tumorigenesis tumours in lymphadenopathy examinations by analysing histopathology slides with coloured clinical specimens. This is not the first use of AI to undertake histological assessment, although it is intriguing that this system may detect worrisome spots in the pathological examination that are indistinguishable to the naked eye. On two datasets, LYNA was demonstrated to properly categorise samples as malignant or benign 99.9% of the time. In addition, when clinicians used LYNA in combination with their normal examination of coloured clinical specimens, the median image reviewing duration was cut in half (Van et al., 2022).

Additional imaging-based methods have successfully demonstrated a comparable capacity to improve clinician performance. In the near future, clinicians could utilise such technologies to double-check their diagnosis and understand medical information more quickly without compromising performance. However, in the long run, administration-authorised processes might be able to work autonomously in the clinic, freeing up clinicians to concentrate on instances that computers can't handle. Both LYNA and DLAD are prime examples of systems which aid doctor findings of healthful and unhealthy examples by emphasising visual characteristics that should be examined in greater detail. These works reveal the potential benefits of strategies in healthcare (Jiang et al., 2021).

As of now, medical strategies have demonstrated significant benefits both for physicians and patients. Nonetheless, regulating these systems is difficult. The FDA of the United States has authorised additional methods, although there are presently no uniform clearance standards. In addition, the individuals who develop strategies to be used in the hospital are not necessarily the physicians who treat patients; hence, scientists may need to understand more about healthcare, whilst physicians may have to understand more about activities for which a particular algorithm is or is not well-suited. Even though AI could assist with diagnostics and core clinical duties, it is difficult to conceive of robotic brain surgeries, for instance, in which surgeons must often adjust their strategy on the fly after observing the client. In this and other respects, the potential of AI in medicine presently exceeds its capacity for patient care. Clarified FDA regulations, on the other hand, may assist specify requirements for methods and lead to an increase in the medical deployment of methods (Vellido, 2019).

Additionally, the FDA has severe requirements for investigational approval that demand clear accountability regarding modern science. Several systems depend on highly complex, challenging to decipher maths, frequently referred to as a "black box," to go from the inputs to the output. The possibility that the FDA will authorise a research utilising AI may be impacted by the difficulty to "unpack the black box" and explain how a system functions. Possibly. It seems sense that scientists, businesses, and businessmen could be reluctant to reveal their exclusive techniques to the community for afraid of ruining revenue if their concepts are appropriated and reinforced by the others. The uncertainty around algorithmic specifics might be reduced if patent laws alter from their current condition, in which an approach is theoretically only patentable if it is a component of an underlying hardware. In either case, enhancing openness in the near term is essential to prevent erroneous handling or classification of patient information, making it possible to more easily assess if an approach would be reliable enough to be used in a clinical setting (Laptev et al., 2022).

In contrast to challenges with FDA approval, consumers' confidence and authorisation of AI algorithms could also present challenges. Individuals may well not consent to having an approach utilised to assist with their healthcare issues if those authorising it do not have a clear knowledge of how it operates. If given the option, would people choose to have their illnesses misinterpreted by a person or a computer, even if the method is more accurate than a doctor? Several people find it difficult to respond to this issue, but it essentially comes down to having faith in an algorithm's judgement. The format of the information utilised as input is vital for proper functioning and impacts the accuracy of the decisions. Computers can produce false conclusions when given misleading information. It is entirely conceivable for those who develop algorithms to be unaware that the information they utilise is false till it is too late and their program has resulted in medical malpractice. By having a thorough knowledge of the information and the procedures required to use the data effectively in the algorithms, both doctors and developers could refrain from making this error. It will be less difficult for a computer to train to make errors if there are connections made between scientists developing the algorithms and the physicians who are familiar with the particulars of the clinical evidence (Aung et al., 2021).

Physicians must comprehend the constraints of methods, and developers must comprehend clinical evidence, in order to develop methodologies that are clinically applicable. It may be vital for corporations to reveal the inner workings of their program so that a wider audience may evaluate the approaches and identify error causes that might harm patient care. Due to the absence of a defined procedure for medical clearance, it appears that computers are still a long way from autonomously operating in hospitals. Defining the qualifications needed for a method to be regarded adequately precise for the health centre, trying to address the prospective sources of errors in the algorithm's decision making, and being transparent as to where a computation excels and when it keeps failing might pave the way for the public's acceptance of methodologies to replace physician in certain activities. In order to uniformly improve the precision and efficacy of health systems for varied illnesses, it is worthwhile to attempt to overcome these obstacles (Chan & Zery, 2019).

Figure 1.3 shows the various applications of AI in healthcare.

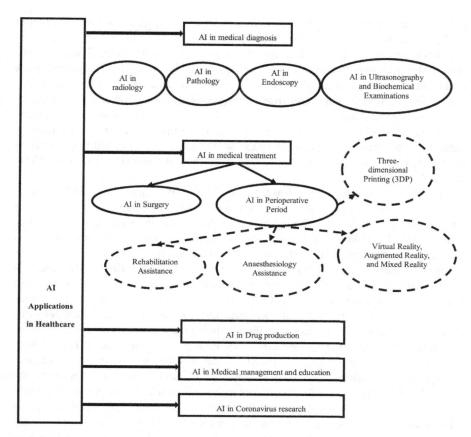

FIGURE 1.3 AI Applications in Healthcare.

1.3 DIAGNOSTIC AI

Whenever a clinician diagnoses a client with a particular ailment using AI, the time necessary to make a diagnostic can be drastically decreased, and diagnostics productivity could be dramatically enhanced. By analysing medical evidence from diagnostic imaging (such as X-ray, CT, and MRI), psychopathic, laparoscopic, and metabolic inspections for associated human body pointers, AI can yield results rapidly and start changing the ineffectual traditional medical prototype, which cannot provide timely and precise findings, particularly for complex diagnoses. In fact, because AI can resolve issues so quickly, clinicians may create a more thoughtful and appropriate therapy strategy based on the patient's situation (Tursunbayeva & Renkema, 2022).

1.3.1 RADIOLOGICAL AI

Radiology currently plays a significant role in the prevention of practically all illnesses as a technical and emotional foundation for clinical diagnosis. The need for radiographic diagnoses is growing rapidly every year, but it takes time to develop medical expertise, and the number of physicians with radiation treatment expertise is only rising gradually. High occupational stress and levels of misdiagnosis are seen as a result of the growing imbalance among the supply and demand of general practitioners in this area. Finding additional methods to control the emergency issue, such as AI, is therefore extremely important from a practical standpoint. Several uses of AI in diagnostic imaging have been discovered in recent years. For instance, Francesco discovered a novel method with a great sensitivity for the advance assessment, quick detection, and grading of retina illnesses via AI deep learning research (Sorrentino et al., 2020), which was further supported by additional study (Heydon et al., 2021; Xie et al., 2020).

Furthermore, developed an AI-aided analytic (CADx) framework solely on medical measurements to categorise normal and cancerous lung nodules by examining 243 patients with corroborated lung nodules. The authors demonstrated the viability of the CADx framework for precisely differentiating the essence of lung nodules as well as the workability of the scheme for initial and subtle diagnosis of lung cancer. To further enhance the study's validity of lung nodule categorisation, the researcher's team combined the quantifiable imagery (QI) characteristics (AI technology) and plasma levels biochemical markers of 173 patients with lung growths. They eventually obtained an encouraging result, demonstrating improved CADx system performance through using QI characteristics than by using only serum biological markers

There is also research for the efficiency of clinical doctors and AI. Like how intelligent technology-enhanced breast cancer screening outperformed clinical radiologists, exhibiting a lower risk of misdiagnosis and an 88% decrease in workload (McKinney et al., 2020); furthermore, the automated preselect feature greatly decreased the demands on doctors. Additionally, it has been demonstrated that an AI system performs superior to human observers when analysing MRI data, greatly enhancing the positive predictive value of mild inflammatory identification in preclinical inflammatory arthritis (Stoel, 2019).

1.3.2   AI in Pathology

The foundation for diagnosing cancers as well as other abnormalities is histology. Whole-slide imagery technology has evolved with the advancement of pathological assess the condition and software applications, becoming a clinical screening approach in pathology research. Nevertheless, there remain issues with how to quickly and mechanically assess and determine an appropriate diagnostic from the practical pathological imagery, necessitating an immediate fix. A new approach has been suggested by several academics who claim that "AI represents the next stage and potential of precision pathology" (Acs et al., 2020; Allen, 2019; Bera et al., 2019; Serag et al., 2019). The use of AI in forecasting diseases has shown promising results. The procedures of pathological imagery classification, tumour diagnosis, and metastatic assessment have been enhanced during the data analysis with the addition of the AI system, and the task is completed in a better quality and smaller timeframe (Wang et al., 2019).

Additionally, studies have demonstrated that the AI system can diagnose abnormal pictures more accurately than expert medical pathologists in several instances (Komura & Ishiwaka, 2019). For instance, Hart has used a convolutional neural network to accurately discriminate between Spitz and traditional melanomas tumours, two distinct disease kinds (Hart et al., 2019). Utilising AI deep learning, Kosaraju has also put forth a novel multitask system that can concurrently capture cross-patch imagery for pathological image processing. The novel model beat existing cutting-edge AI techniques after analysing the pathological imagery of well-, moderately-, and poorly differentiated stomach cancer (Kosaraju et al., 2020). Additionally, Coudray verified that the DL model would, with a high precision of 97%, identify six genetic alterations linked to cancer and help medical professionals identify the subtypes and gene mutation in diagnosis of cancer (Coudray et al., 2018). Additionally, epithelium malignancies, lung disease, basal cell carcinoma, and glomerular have all been diagnosed using AI (Iizuka et al., 2020; Kanavati et al., 2020; Wang et al., 2019; Jiang et al., 2020; Bueno et al., 2020). These findings highlight how useful AI technology is when used in histology.

1.3.3 AI in Endoscopy

With the use of AI technology, there has been significant improvements in endoscope identification which have altered the conventional model and increased efficiency. Certain specialists claim that AI technology could successfully support the endoscopy-based diagnosis of tumours and colon abnormalities, in addition to gastrointestinal and oesophagus malignancies (Namikawa et al., 2020). Colonoscopy, according to Gulat and Emmanuel, is a promising technique for AI advancement. After DL, the AI system can significantly improve the prognosis of stomach and genitourinary illnesses, such as Barrett's oesophagus, villous melanoma, and oesophageal cancer, by speeding up the sensing process and increasing analytic precision (Gulati et al., 2020). The findings indicate that endoscopy in combination with the innovative AI algorithm had a greater sensitivity and more precise translation of the intestinal tumours than with the conventional model (Hwang et al., 2020). Additionally, some researchers gathered 7556 image data by colonoscopy and then analysed them using AI systems to achieve a useful neural network technique to necessarily identify intestine injuries. There is undoubtedly a potential for these technological advances as more and more studies (He et al., 2019; Chahal & Byrne, 2020; Sharma et al., 2020) support the viability of AI with endoscopy in the classification and diagnosis of infectious illnesses.

1.3.4 AI in Ultrasound and Biological Tests

According to what was stated previously, the deployment of AI technology improves the accuracy of tomography and metabolic testing. While physicians have indeed utilised image-based computer-aided diagnosis (CAD) systems to ultrasound diagnostics, the effectiveness is highly dependent on the classification and detection techniques. In conjunction with AI technology, approaches have evolved significantly.

Nguyen, for instance, has suggested a novel echocardiography imagery analytical technique centered on AI that has effectively improved the identification of thyroid nodules (Nguyen et al., 2019). Other researchers have also proven that AI may enhance the conventional echocardiography identification of cancers in the throat, chest, bronchus as well as other obstetrics and gynaecologic diseases, with a significant level of efficacy and precision (Nguyen et al., 2020; Sun et al., 2020; Chen et al., 2020; Fujioka et al., 2019; Chen et al., 2019).

Additionally, AI has made considerable strides in the identification of clinical diseases following a thorough training study of diagnostic test statistics based on the paradigm of big-data assessment. Abelson has created an ML prediction method that aids in initial detection and surveillance by using deep sequencing to examine markers which are frequently altered in chronic leukaemia (ML) and big electronic health records databases analytics (Abelson et al., 2018). Furthermore, Sun used an AI method to analyse the CT imageries and RNA sequencing information from patients with haematological malignancies in order to correctly forecast the based biometric signal and medical results in anti-PD-1/PDL1 chemotherapy (Sun et al., 2018). AI also helps with the identification of Noonan syndrome, a prevalent inherited condition, particularly in unusual people, by examining the genes and associated disease manifestations (Li et al., 2019). Additionally, the prognosis and identification of initial asthma were significantly aided by using AI deep learning to create a multilayer perceptron of numerous medical trials, including pulmonary function tests, bronchial trials, as well as some molecular techniques from 556 individuals (Tomita et al., 2019).

Researchers who adopt the opposing view, nevertheless, contend that while AI has a role in medical assessment, its true impact could develop slowly over time and won't overtake physicians in the near future (Demircioglu, 2019). Nevertheless, AI plus diagnosis has emerged as the overall trend at this point and will keep on growing.

1.4   AI IN MEDICAL TREATMENT

1.4.1   AI IN SURGERY

The clinical AI system represents the pinnacle of AI's accomplishments and its most important clinical use to date. Twenty years ago, robots like the PUMA-560, Probot, AESOP, Robodoc, and Acrobot (Jakopec et al., 2001; Cowley, 1992; Stefano, 2017) served as useful surgical adjuncts. However, previous therapeutic devices required human oversight and intervention, and thus they amounted to little more than a more adaptable version of a standard scalpel.

With the advancement of AI technology, the notion of a therapeutic model that uses AI has emerged. The Da Vinci surgical AI system represents the most innovative product of this theory in the modern period. The introduction of the Da Vinci medical machine as a significant innovation unique in history of mankind made medical intervention less intrusive, with the benefits of a sharper vision, more precise and simple treatment, and even remote control. This inventive idea enables previously challenging minimally invasive approaches to be used to execute complex surgical procedures. The Da Vinci surgical AI system consists of three components: the surgeon interface, the robotic operating system, and the imaging system. The FDA of the USA authorised the use of the Da Vinci surgical equipment in actual operations in 2000. This AI system transformed the conventional treatment model. With the implementation of the Da Vinci surgical AI system, for instance, thyroid procedure was enhanced in aspects of postoperative symptom control and speech outcomes (Tae, 2020); mesial procedure was enhanced in terms of precision and safety (Stefanelli et al., 2020); intestinal, renal, and prostate gland surgical intervention was enhanced as evidenced by an elevated surgery positive outcome but a low attrition rate (Lenfant et al. 2020; Winder et al., 2020; Jones et al., 2020); and respiratory cancer surgery was advantageous for patients with regard to postoperative recovery (Wang et al., 2020).

The primary distinguishing feature of AI surgical technologies, relative to conventional surgical processes, is AI, which indicates that the surgery processes have evolved to a intelligent state. Assisted by AI techniques such as DL, histologic analysis in the course of surgery is now a reality, allowing for fast incisal edge pathologic investigation and genuine cell biopsies (Zuo & Yang, 2017). Utilising DL, the AI method could also self-deduce centred on the plentiful experimentations from diagnostic physicians and recreate clinical digitised data by posting the surgical programme to an AI surgical scheme to smartly aid the surgical procedure, such as minimally invasive resection scope composition, postoperative endocrine remnant quantity health coverage, and prognostication of lymphatic system with potentially positive metastatic disease (Navarrete & Hashimoto, 2020). The operative strategy and procedures are dependent on not only physicians but additionally on a programme employing an artificial system (Samareh, 2019). Even though AI surgical systems have acquired substantial awareness at the present phase, they still require some manual oversight. Nevertheless, this point will be expanded farther and become a potential powerhouse, one and day the entirety of intelligence would be realised.

1.4.2 Perioperative AI

The perioperative phase encompasses the time between the patient getting surgical intervention and completing basic rehabilitation; it is divided into three areas: preoperative planning, surgery phase, as well as postoperative healing process. There are numerous breakthroughs with the deployment of AI technologies throughout the entire perioperative period.

1.4.2.1 Three-dimensional Printing (3DP)

3DP is a system that incorporates AI in certain of its operations. It is a form of fast prototyping method that employs powdery metals or other sticky biomolecules to build items layer-by-layer from digitised prototypes (generated from CT or MRI information with AI expertise). Integrating medical imagery into automation such as MIMICS. After the user has arbitrarily selected the regions of interest, the program can produce a basic digital three-dimensional reconstruction for publishing via algorithmic evaluation. Even though further analysis by people might well be required at the present phase, we anticipate that one day it will attain total cognition. In the case of complicated internal wounds or broken bones, for instance, it is challenging for medical surgeons to identify the crucial location using conventional investigative techniques throughout preoperative treatment. Nevertheless, when paired with the early phase of 3DP technology—Model Printing—surgeons can handle the 1:1 actual model of the wounded portion reconstructed from the actual CT scanning information, obtain more graphic and intuitive data, and even practise a virtual operation on the model in preparation (Tejo et al., 2020). In vascular and cardiac operations, the 3DP system enables a patient-specific representation that recognises complicated physiology and facilitates damage localisation, providing preventive, and therapeutic relationships, according to researchers (Wang et al., 2020).

Moreover, numerous studies have demonstrated that 3D printing contributes in the presurgical preparation for oral procedure, arthroscopy, surgical intervention, diagnostic and therapeutic surgical intervention, and certain tumour treatments, from boosting preoperative scheduling to boosting the operator's faith (Nikoyan & Patel, 2020; Skelley et al., 2019; Yamaguchi & Hsu, 2019; Bangeas et al., 2019). In its next phase, Operating Guidance, 3D printing performs a critical function not just in preoperative planning and throughout the operation time. In a surgical procedure, there are constantly requirements for intermaxillary fixation and having to cut for orthoses or tumour resection, so complications could emerge, like how to ascertain the optimal angular position and position of fixing for high productivity and where the sharp tool ought to be to be preserved as much healthy cells as potential. Using the preliminary diagnostic information, 3D printing may generate a personalised surgical guideline and a surgical model. Pedicle screw placement was significantly safer and simpler using 3DP templates guidance in surgical treatment than conventional techniques; the danger of adjacent endovascular injury was successfully reduced, and radiation exposure also was lowered (Feng et al., 2018; Kashyap et al., 2018).

Furthermore, uses of 3DP templates guidance in inversion procedures for tibial abnormalities and total joint replacement have indeed been documented; the procedure duration and effectiveness were improved significantly with 3DP technologies compared to conventional approaches (Corona et al., 2018; Sun et al., 2020; Zhou et al., 2020). As for tumour removal, 3DP software allows to accurately find and verify the state of the art in skeletal surgical resection, resulting in a satisfying postoperative outcome (Park et al., 2021), while also minimising the danger of crucial structural harm and conserving additional normal tissues (Gomez et al., 2019). In addition, the most recent step of 3D printing technology, bodily implants, miraculously reconstructs human tissue with biological elements, such as scaffolding elements, functioning organisms, and bioactive components. After printing and sterilisation, the device can be utilised in operation to repair defective or damaged biological material caused by a variety of causes. For example, the implementation of polymeric materials, bioceramics, and composite materials as biomaterial that publish personalised skeletal overhead cranes has been found to improve minimally invasive efficacy and patient preferences; within and between people treated huge mandible deficiency repair surgery, the faulty material was completely fixed using 3D printing technology (Salah et al., 2020).

In surgical treatment, 3DP technologies paired with the mirror-replication approach offers a solution to the unsolvable problem of cranium and limb skeleton abnormalities that previously required marrow grafting (a treatment with several risks) to re-establish bones (Alkhaibary et al., 2020; Vidal et al., 2020; Xing et al., 2020). Likewise, 3DP substitution has been extensively used in neurological illness, muscle tissue re-establishment, prosthesis, heart valve replacements, and genital disorders by urologists and gynaecologists (Rey et al., 2020; Boso et al., 2020; Ettinger & Windhagen, 2020; Levin et al., 2020; Farmer et al., 2020). In addition, 3DP technology will shortly achieve its aim of producing whole, operational, and living organs, paving the way for the subsequent phase of Limb Biotechnology (Edgar et al., 2020).

1.4.2.2 Virtual Reality, Augmented Reality, and Mixed Reality

The digitalisation hologram imaging techniques of virtual reality (VR), augmented reality (AR), and mixed reality (MR) are comparable to 3DP in that they partially use AI techniques to rebuild medical evidence throughout their operations. A pure virtualised image formed by a sophisticated computer program is what is known as VR, and it might give physicians the chance to practise utilising a virtual network without worrying about the potential negative effects of an unsuccessful surgery (Mirchi et al., 2020; Sadeghi et al., 2020). VR, though, can't be used in actual treatment because it lacks practical experience (Fertleman et al., 2018). In principle, AR differs from VR due to its focus on the actual world. AR is a combination of artificially combined data and the physical system. AR technology could legitimately support treatment intraoperatively or intraoperatively by detecting the intricate anatomical characteristics and assisting with navigation mostly during operation after converting patient information and virtually reasserting the crucial region (Creighton et al., 2020; Gibby et al., 2020). There are still restrictions during operation, though, because AR guidance technology requires cumbersome hardware (Hu et al., 2019). The issue has been successfully solved thanks to the convergence of VR and MR, which dissolves the distinction among simulated reality and actuality with the advent of the most innovative online hologram computed tomography.

The three characteristics of MR are accurate identification, actual contact, and closed integration of virtuality with reality (Goo et al., 2020). The diagnostic physician could indeed delve oneself in the blended operating universe and develop a improved treatment plan in the MR scheme, that is made up of similarly portable equipment (such as a portable MR gadget, the Hololens, and the most recent software technical manufacturing (Salmas et al., 2020), a real-time interactive placement, and visually compelling perspectives (Wu et al., 2018). Additionally, the doctor-patient relationship is enhanced. Because of these benefits, this novel technique has been used for perioperative guiding aid in a range of sectors, including spinal, orthopaedic, hepatic, renal, as well as cranial operations, lowering the operating frequency and enhancing the precision and efficiency of operations (Chytas et al., 2021; Zeiger et al., 2020; Wu et al., 2018; Yoshida et al., 2020). As data could be communicated in live time utilising internet chatting systems, MR has even satisfied the criteria for telehealth, which is crucial for providing healthcare in rural and isolated places (Rojas et al., 2020). With the exception of more precise steering, MR still has certain benefits over 3DP in terms of speed since 3DP manufacturing can require several hours to produce (Hu et al., 2019). Additionally, MR software can help with ordinary workout and postoperatively therapy (Held et al., 2020; Chen et al., 2020).

1.4.2.3 Anaesthesiology Support

Throughout the postoperative period, AI technology has also been broadly applied in anaesthesia. Anaesthesia is an integral aspect of the medical operation that facilitates a flawless surgery; nonetheless, it is fraught with numerous dangers and problems. Coupled with the use of AI technology, six aspects have gained considerable exposure and focus: (1) anaesthetic level tracking; (2) anaesthetic regulation; (3) adverse event detection; (4) ultrasonography support; (5) chiropractic adjustments; and (6) surgical strategic planning (Edgar et al., 2020). Artificial intelligence technology boosts the safety of surveillance, administration, and postoperatively care, providing exciting advancements to anaesthesia (Hashimoto et al., 2020; Seger et al., 2020; Kamdar & Jalilian, 2020).

1.4.2.4 Rehabilitation Support

AI technology also has a significant impact on postoperative rehabilitation and the healing process. For instance, the use of AI wireless sensors in the intensive care unit (ICU) can efficiently collect patient data, decrease false alarms, and ease ICU problems (Poncette et al., 2020). There are numerous innovative tools introduced to the nursing profession with the gradual diversification of AI technology (Angehrn et al., 2020). The AI-based medical equipment can speed up the process while also addressing the needs of rehabilitation for patients who are recovering (Dai et al., 2020). Furthermore, the use of AI robots has sped up limb physiotherapy in sophisticated anthropopathic action direction and assisted patients in recovering to a greater extent (Averta et al., 2020; Zhao et al., 2020). Additionally, AI technology has been utilised to pursue development and supervise health, which might be useful for the maintenance of individuals who have been released (Vos et al., 2020; Ramezani et al., 2019).

1.5 AI IN DRUG CREATION

In the classic model, the manufacturing of prescription medications necessitates a extended and complex period of time, which includes efficient target research, medication constituent strategy, performance analysis, diagnostic tests, experimenting, and advancement; consequently, even after a lengthy time of investigation, experimental medicines may not operate as anticipated. In recent times, though, the growth of AI has altered the natural drug companies in healthcare and accelerated the identification and manufacturing of new medicines (Bajorath et al., 2020, Brown et al., 2020). In addition, as a result of the continuous maturation of AI-generation medications, both the originality and effectiveness of treatment have achieved greater levels (Zhavoronkov, 2020). For example, the integration of AI estimation techniques with vaccine design has effectively expedited investigational procedures and reduced the costs and duration of research and development (Russo et al., 2020). Drug research driven by DL technologies could target proteins as intended, and that was previously unavailable (Fernandez, 2020). As a result of the powerful analytical contribution and computerised knowledge capabilities of AI technology, the design and production of cancer treatments have been significantly improved in terms of their therapeutic efficacy (Liang et al., 2020). In particular, research into AI-assisted genomics methods and tools has presented a lot of promise for specific molecular medicinal therapy (Takakusagi et al., 2020). In addition, the above-mentioned 3DP technique has led to significant advances in medication manufacture. 3DP allows for the choice of pharmaceutical size, form, and the mixture of various medicinal ingredients, which might be more practical for therapeutic trials (Awad et al., 2020). Even the layering and percentage factors in pill coatings, medicinal discharge timings, and patterns can be predesigned using 3DP technology, resulting in enhanced therapeutic benefits (Pandey et al., 2020; Tsintavi et al., 2020).

1.6 AI APPLICATIONS IN HEALTHCARE ADMINISTRATION AND ACADEMIA

In the classical method, medical management depends on the general design of the administration office, and there are frequently managerial flaws and drawbacks such as the arbitrary allocation of healthcare resources. The regulation of AI technology has rapidly altered the way things are done. In order to anticipate the exact latency in the emergency room of their health centre, a few researchers built a forecasting model and analysed the repository of patient hospital stays using LSTM models. This improved greatly the healthcare effectiveness, patients' subjective experiences, and supported the redistribution of medical resources (Cheng & Kuo A, 2020). The overall result of hospital stays was decreased by 7%, the ideal multitude of hospital wards was chosen, and hospital funds and necessary inputs were optimised, according to studies that used 10 AI algorithms to analyse information on patient hospital-stay times, routes to hospitals, along with climatical and time-based components (Nas & Koyuncu, 2019). Additionally, a practical forecasting model centred on synthetic neural networks effectively and comparably forecasted the readmissions, assisting in treatment planning and enhancing medical administration (Saab et al., 2020). In summary, AI technology has made it easier to administer hospitals, provide tailored clinical treatment, and advise patients.

The training of doctors is the future and the promise of advancements in medicine, but it is a lengthy and challenging process because of the vast and sophisticated technical necessary knowledge. If medical students merely study textbooks and samples, their progress would be hampered. Medical students' training has become deeper and more colourful as a result of the varied applications of AI technology. Students' knowledge and comprehension have increased thanks to AI-based problem-based learning, which has also increased their understanding of medical symptoms (Wu et al., 2020). Medical students' competence and trust have improved as a result of the integration of researching operation with an AI system (Yang & Shulruf, 2019). Additionally, a novel instructional tool providing external guidance has been developed by the AI model-centred clinical training program, which integrates AI and simulations for learning surgical skills (Mirchi et al., 2020). AI technology has made it possible to analyse students' psychological health and educational achievement in addition to helping with education, which allows institutions to be more aware of their pupils' issues more quickly (Dekker et al., 2020). Additionally, the 3DP and MR technologies could give medical trainees more engaging learning possibilities that are not possible with traditional textbook reading. Three-dimensional restoration differs from two texts thanks to sophisticated methods; as a result, learners can use the 3DP healthcare model to study 3D structural structures and even practise procedures on the model to advance their medical abilities (Bertin et al., 2020; Bohl et al., 2020). Presently, 3DP- or MR-based support techniques have been extensively used in medical education. These approaches could aid students in understanding the social structure in a more interesting way including any size or layer which they monitor, giving simulative surgical education without hazards (Sappenfield et al., 2018).

1.7 USING AI FOR COVID STUDIES

The novel coronavirus illness 2019 (COVID-19) epidemic introduced hazards to the globe at the end of 2019. In the worldwide calamity, public health, security, the advancement of modern civilisation, as well as the world economy were badly impacted, and numerous individuals perished. Fortunately, considerable growth was done in the diagnosis and therapy of COVID-19 by employing several innovative diagnostic procedures and fully advanced technology, particularly AI models. To regulate the spread of COVID-19, AI was utilised as a substitute for social intellect to discuss initial identification and diagnosis, patient management, investigations, prognostication of cases and death rates, advancement of medications and vaccines, decrease of healthcare workload, and preventative medicine (Vaishya et al., 2020). Since the quantitative measurements and translation of respiratory lung CT information could be assessed precisely and effectively during in the fight against COVID-19, Zhang created a novel tool based on DL to analyse the CT information from patients and deduced that the lower right hindbrain of the respiratory system is the region with the highest incidence of COVID-19 bronchitis (Zhang et al., 2020). Moreover, Aikaterini introduced an AI algorithm for the processing of COVID-19 CT images, and their results suggested that the procedure might facilitate early identification and hospital attention (Sakagianni et al., 2020). In addition, Tivani offered point-of-care diagnosis and treatment that incorporated imaging, histology, and AI to assist in the diagnosis of COVID-19 (Zhang et al., 2020). Moreover, Sweta used a drug-repositioning technique to conduct a rapid intelligent testing for prospective drugs to treat COVID-19; this group was able to identify possibly beneficial substances employing both AI- and pharmacology-centered methodologies, illustrating that this technique might be beneficial for COVID-19 drug design and study. This strategy has also been corroborated by the other researchers, who developed a system using AI learning and modelling techniques to find the medications on the marketplace with the ability to treat COVID-19; consequently, they identified and over 80 pharmaceuticals with strong potential (Ke et al., 2020). In addition, there has been extensive study on AI method support, which has accelerated the creation of COVID-19 vaccinations (Kim et al., 2020; Arash et al., 2020; Elaziz et al., 2020).

With adequate AI-centered technologies, the processes of initial alert, assessment, medicine development, and healthcare management throughout the battle over COVID-19 would be efficiently ensured (Mali & Pratap, 2021), and the epidemic will soon be defeated.

TABLE 1.1
AI Applications and Techniques in Healthcare

AI Application	Reference	Task	AI technique
Radiology	(Sorrentino et al., 2020; Heydon et al., 2021; Xie et al., 2020)	Early detection, prompt treatment, and classification of retinal disorders	Deep neural network
	(Rodriguez et al., 2019a; McKinney et al., 2020; Rodriguez et al., 2019b)	Identify cancerous and harmless pulmonary nodules Detection of Breast Cancer	AI-aided diagnostic (CADx) CNN
	(Stoel, 2019)	premature arthritis diagnosis	CNN
Pathology	(Acs et al., 2020; Allen, 2019; Bera et al., 2019; Serag et al., 2019; Wang et al., 2019; Komura & Ishiwaka, 2019)	tumour detection, metastasis assessment, and abnormal segmentation techniques	Deep learning
	(Hart et al., 2019)	separate Spitz tumours from ordinary melanomas tumors	CNN
	(Kosaraju et al., 2020)	esophageal cancer diagnosis	multitask-based inception V3 model
	(Coudray et al., 2018)	identify the subtype and genetic mutations while diagnosing malignancy	Inception V3 model
	(Iizuka et al., 2020; Kanavati et al., 2020; Wang et al., 2019; Jiang et al., 2020; Bueno et al., 2020)	Endothelial tumours, lung disease, basal cell carcinoma, and glomerulosclerosis were identified.	CNN, RNN, EfficientNet-B3. Cascade of 2 Deep neural networks.
Endoscopy	(Namikawa et al., 2020)	cancer screening (including colorectal, stomach, and gastroesophageal)	CNN
	(Gulati et al., 2020)	diagnostic testing for disorders of the digestive tract and abdomen, such as colon cancer, basal cell carcinoma, and Barrett's esophageal	SVM, CNN, Decision tree

(continued)

TABLE 1.1 (Continued)
AI Applications and Techniques in Healthcare

AI Application	Reference	Task	AI technique
	(Hwang et al., 2020)	determining the exact location of the intestinal tumours	CNN
	(He et al., 2019; Chahal & Byrne, 2020; Sharma et al., 2020)	evaluated the level of infiltration of cancerous GI tumours and distinguish between benign and malignant lesions	CNN, "computer-assisted detection (CADe)", "computer-assisted diagnosis (CADx)"
Ultrasonography and Biochemical Examinations	(Nguyen et al.,2019)	categorization of thyroid nodules	CNN, Resnet
	(He et al., 2019; Chahal et al., 2020; Sharma et al., 2020; Nguyen et al., 2019; Nguyen et al., 2020; Sun et al., 2020; Chen et al., 2020; Fujioka et al., 2019; Chen et al., 2019)	Pancreatic, ovarian, bronchial, pectoralis major tendon, and urinary interval cancers, various obstetrics and gynaecological abnormalities were detected with ultrasonography.	CNN ensemble
	(Abelson et al., 2018)	surveillance and identification of chronic lymphocytic leukaemia	Deep neural network
	(Sun et al., 2018)	Anti-PD-1/PDL1 monoclonal antibodies: Estimation of the radiomic signal and diagnostic procedures	technique of elastic-net regularised analysis
	(Li et al., 2019)	diagnosis of Noonan syndrome,	CNN
	(Tomita et al., 2019)	prediction and diagnosis of asthma	DNN
Surgery	(Demircioglu, 2019; Jakopec et al., 2001; Cowley, 1992; Stefano, 2017)	Surgical systems	PUMA-560, Probot, AESOP, Robodoc, and Acrobo
	(Tae, 2020; Stefanelli et al., 2020)	thyroid surgery	Da Vinci surgical AI system
	(Lenfant et al., 2020; Jones et al., 2020)	gastric, nephritic, and prostatic surgery	
	(Wang et al., 2020)	Lung cancer surgery	
	(Zuo & Yang, 2017)	real-time tissue biopsy	

Application	References	Description	AI methods
	(Navarrete & Hashimoto, 2020)	operative resection spectrum creation, perioperative tissue reserve area coverage, plus lymphadenopathy prognosis	Multi-Support Vector Regression, SVM, CNN /DNN shared design
3DP	(Samareh et al., 2019; Tejo et al., 2019; Wang et al., 2020)	generate a fundamental digital three-dimensional restoration for printing based on algorithmic evaluation	MIMICS
Virtual Reality	(Nikoyan & Patel, 2020; Yamaguchi, & Hsu, 2019)	The effectiveness of a posterior cervical surgical removal in virtual reality	ANN
Augmented Reality	(Bangeas et al., 2019; Feng et al., 2018; Kashyap et al., 2018)	Recognition of the complicated structural forms as well as navigation throughout surgery	Digital holographic imaging technology
Mixed Reality	(Corona et al., 2018; Sun et al., 2020; Zhou et al., 2020; Park et al., 2021; Gomez et al., 2019; Salah et al., 2020; Alkhaibary et al., 2020; Vidal et al., 2020; Xing et al., 2020; Rey et al., 2020; Boso et al., 2020)	cervical, musculoskeletal, invasive procedures involving the liver, kidneys, and cerebral, as well as subsequent rehab and periodic training	An MR headset that can be worn, known as the Hololens, is the most modern technical innovation from Microsoft.
Anaesthesiology Assistance	(Ettinger & Windhagen, H., 2020; Levin et al., 2020; Farmer et al., 2020)	anaesthetic level tracking, anaesthesia regulation, incident forecasting, ultrasonography support, medication management, and neurosurgical administration	ANN, Fuzzy logic, ML, Computer vision
Rehabilitation assistance	(Edgar et al., 2020; Mirchi et al., 2020; Sadeghi et al., 2020; Fertleman et al., 2018; Creighton et al., 2020; Gibby et al., 2020; Hu et al., 2019)	patient data collection, reduction of pseudo alarms, patient rehabilitation, advancement tracking, and wellbeing monitoring.	SVM, ANN, Principal component analysis
Drug production	(Goo et al., 2020; Salmas et al., 2020; Wu et al., 2018a; Chytas et al., 2021; Zeiger et al., 2020; Wu et al., 2018b; Yoshida et al., 2020; Rojas et al., 2020; Held et al., 2020; Chen et al., 2020)	New drug discovery and assembly, target proteins, pattern and manufacture of malignancy medicines, small molecule drug therapy, choice of medication size, appearance, and blend of different pharmaceutical ingredients,	Random Forest, Gradient Boosted Trees, or Gaussian Processes, DNNs, CNNs, or RNNs

(continued)

TABLE 1.1 (Continued)
AI Applications and Techniques in Healthcare

AI Application	Reference	Task	AI technique
Medical management and education	(Hashimoto et al., 2020; Seger & Cannesson, 2020; Kamdar & Jalilian, 2020; Poncette et al., 2020; Angehrn et al., 2020; Dai et al., 2020; Averta et al., 2020; Zhao et al., 2020; Ramezani et al., 2019; Bajorath et al., 2020)	Likelihood of expecting periods in hospital, reduce the average hospitalisation time, patient therapy, clinic administration, health source allotment, and finally customised medical treatment	Long-short-term memory (LSTM), ANN, RNN, Explainable AI
COVID research	(Zhavoronkov, 2020; Russo et al., 2020; Fernandez 2020; Liang et al., 2020; Takakusagi et al., 2020; Awad et al., 2020; Pandey et al., 2020; Tsintavi et al., 2020; Cheng et al., 2020)	Timely detection and identification, therapeutic targets, interaction tracking, incidence and fatality prognosis, medication and vaccine research, professional burden decrease, and preventative medicine	DNN, revised 3D CNN and an integrated V-Net with restricted access forms, CNN, RNN, Deep Belief Networks

1.8 CONCLUSION AND FUTURE SCOPE

AI methods is an advanced invention that evolves with the modern world; consequently, it is a natural outcome of the progress of science and expertise over time. In the history of mankind, there have been two manufacturing innovations: the steam uprising and the electric innovation, each of which significantly altered human life and advanced civilisation. The current technological and scientific transformation, which includes AI technology, has already demonstrated an unstoppable pattern has exploded in popularity like a wildfire. The conventional healthcare setting in the medicine field has been substantially altered thanks to novel AI methods; patient prognosis utilising radiographic, pathogenic, laparoscopic, ultrasonographic, and physiochemical tests has also been successfully endorsed with a greater level of precision and reduced human volume of work. Despite superior medical results, the medical treatment provided throughout the intraoperative phase, which includes preoperative planning, the operating phase, and the postoperative healing process, has. been greatly improved. Additionally, AI technology has been instrumental in reshaping the fields of healthcare administration, professional examinations, and drug development. In this chapter, different AI methods along with real-world applications were covered. The era of AI has arrived, and we anticipate that this new revolution will change healthcare in a way that has never been seen before.

CONFLICT OF INTEREST

The authors have no conflicts of interest to declare. All co-authors have seen and agree with the contents of the manuscript and there is no financial interest to report. We certify that the submission is original work and is not under review at any other publication.

REFERENCES

Abelson, S., Collord, G., Ng, S., et al. (2018). Prediction of acute myeloid leukemia risk in healthy individuals. *Nature*, 559(7714), 400–404.

Acs, B., Rantalainen, M., & Hartman, J. (2020). Artificial intelligence as the next step towards precision pathology. *Journal of Internal Medicine*, 288(1), 62–81.

Alkhaibary, A., Alharbi, A., Alnefaie, N., et al. (2020). Cranioplasty: A comprehensive review of the history, materials, surgical aspects, and complications. *World Neurosurgery*, 139, 445–452.

Allen, T. C. (2019). Regulating artificial intelligence for a successful pathology future. *Archives of Pathology & Laboratory Medicine*, 143(10), 1175–1179.

Amisha, Malik, P., Pathania, M., & Rathaur, V. K. (2019). Overview of artificial intelligence in medicine. *Journal of Family Medicine and Primary Care*, 8(7), 2328–2331.

Angehrn, Z., Haldna, L., Zandvliet, A. S., et al. (2020). Artificial Intelligence and machine learning applied at the point of care. *Frontiers in Pharmacology*, 11, 759.

Arash, K. A., Julia, W., Milad, S., et al. (2020). Artificial intelligence for COVID-19 drug discovery and vaccine development. *Frontiers in Artificial Intelligence*, 3, 65.

Aung, Y. Y. M., Wong, D. C. S., & Ting, D. S. W. (2021). The promise of artificial intelligence: A review of the opportunities and challenges of artificial intelligence in healthcare. *British Medical Bulletin*, 139(1), 4–15. https://doi.org/10.1093/bmb/ldab016

Averta, G., Della Santina, C., Valenza, G., et al. (2020). Exploiting upper-limb functional principal components for human-like motion generation of anthropomorphic robots. *Journal of NeuroEngineering and Rehabilitation*, 17(1), 63.

Awad, A., Fina, F., Goyanes, A., et al. (2020). 3D printing: Principles and pharmaceutical applications of selective laser sintering. *International Journal of Pharmaceutics*, 586, 119594.

Bajorath, J., Kearnes, S., Walters, W. P., et al. (2020). Artificial intelligence in drug discovery: Into the great wide open. *Journal of Medicinal Chemistry*, 63(16), 8651–8652.

Bangeas, P., Tsioukas, V., Papadopoulos, V. N., et al. (2019). Role of innovative 3D printing models in the management of hepatobiliary malignancies. *World Journal of Hepatology*, 11(7), 574–585.

Basu, K., Sinha, R., Ong, A., & Basu, T. (2020). Artificial intelligence: How is it changing medical sciences and its future? *Indian Journal of Dermatology*, 65(5), 365–370.

Bera, K., Schalper, K. A., Rimm, D. L., et al. (2019). Artificial intelligence in digital pathology – new tools for diagnosis and precision oncology. *Nature Reviews Clinical Oncology*, 16(11), 703–715.

Bertin, H., Huon, J. F., Praud, M., et al. (2020). Bilateral sagittal split osteotomy training on mandibular 3-dimensional printed models for maxillofacial surgical residents. *British Journal of Oral and Maxillofacial Surgery*, 58(8), 953–958.

Bohl, M. A., McBryan, S., Pais, D., et al. (2020). The living spine model: A biomimetic surgical training and education tool. *Operative Neurosurgery (Hagerstown)*, 19(1), 98–106.

Boso, D., Maghin, E., Carraro, E., et al. (2020). Extracellular matrix-derived hydrogels as biomaterial for different skeletal muscle tissue replacements. *Materials*, 13(11), 2483.

Briganti, G., & Le Moine, O. (2020). Artificial intelligence in medicine: Today and tomorrow. *Frontiers in Medicine*, 7, 27.

Brown, N., Ertl, P., Lewis, R., et al. (2020). Artificial intelligence in chemistry and drug design. *Journal of Computer-Aided Molecular Design*, 34(7), 709–715.

Brufau, S. R., Wyatt, K. D., Boyum, P., Mickelson, M., Moore, M., & Cognetta-Rieke, C. (2019). A lesson in implementation: a pre-post study of providers' experience with artificial intelligence-based clinical decision support. *International Journal of Medical Informatics*, 137, 104072.

Bueno, G., Fernandez, C. M., Gonzalez, L., et al. (2020). Glomerulosclerosis identification in whole slide images using semantic segmentation. *Computer Methods and Programs in Biomedicine*, 184, 105273.

Chahal, D., & Byrne, M. F. (2020). A primer on artificial intelligence and its application to endoscopy. *Gastrointestinal Endoscopy*, 92(4), 813–820.

Chan, K. S., & Zary, N. (2019). Applications and challenges of implementing artificial intelligence in medical education: Integrative review. *JMIR Medical Education*, 5(1), e13930. https://doi.org/10.2196/13930

Chen, C. H., Lee, Y. W., Huang, Y. S., et al. (2019). Computer-aided diagnosis of endobronchial ultrasound images using convolutional neural network. *Computer Methods and Programs in Biomedicine*, 177, 175–182.

Chen, P. J., Penn, I. W., Wei, S. H., et al. (2020). Augmented reality-assisted training with selected Tai-Chi movements improves balance control and increases lower limb muscle strength in older adults: A prospective randomized trial. *Journal of Exercise Science & Fitness*, 18(3), 142–147.

Cheng, N., & Kuo, A. (2020). Using Long Short-Term Memory (LSTM) neural networks to predict emergency department wait time. *Studies in Health Technology and Informatics*, 272, 199–202.

Chytas, D., Chronopoulos, E., Salmas, M., et al. (2021). Comment on: "Intraoperative 3D hologram support with mixed reality techniques in liver surgery." *Annals of Surgery*, 274(6), e761–e762.

Coombs, C., Hislop, D., Taneva, S. K., & Barnard, S. (2020). The strategic impacts of Intelligent Automation for knowledge and service work: An interdisciplinary review. *Journal of Strategic Information Systems*, 29(4), 101600.

Corona, P. S., Vicente, M., Tetsworth, K., et al. (2018). Preliminary results using patient-specific 3D printed models to improve preoperative planning for correction of post-traumatic tibial deformities with circular frames. *Injury*, 49(Suppl 2), S51–S59.

Coudray, N., Ocampo, P. S., Sakellaropoulos, T., et al. (2018). Classification and mutation prediction from non-small cell lung cancer histopathology images using deep learning. *Nature Medicine*, 24(10), 1559–1567.

Cowley, G. (1992). Introducing "Robodoc." A robot finds his calling—in the operating room. *Newsweek*, 120(21), 86.

Creighton, F. X., Unberath, M., Song, T., et al. (2020). Early feasibility studies of augmented reality navigation for lateral skull base surgery. *Otology & Neurotology*, 41(7), 883–888.

Dai, B., Yu, Y., Huang, L., et al. (2020). Application of neural network model in assisting device fitting for low vision patients. *Annals of Translational Medicine*, 8(11), 702.

Dekker, I., De Jong, E. M., Schippers, M. C., et al. (2020). Optimizing students' mental health and academic performance: AI-enhanced life crafting. *Frontiers in Psychology*, 11, 1063.

Demircioglu, A. (2019). Radiomics-AI-based image analysis. *Der Pathologe*, 40(Suppl 3), 271–276.

Dreyer, K., & Allen, B. (2018). Artificial intelligence in health care: Brave new world or golden opportunity? *Journal of the American College of Radiology*, 15(4), 655–657.

Edgar, L., Pu, T., Porter, B., et al. (2020). Regenerative medicine, organ bioengineering, and transplantation. *British Journal of Surgery*, 107(7), 793–800.

Elaziz, M. A., Hosny, K. M., Salah, A., et al. (2020). New machine learning method for image-based diagnosis of COVID-19. *PLOS One*, 15(6), e235187.

Esmaeilzadeh, P. (2020). Use of AI-based tools for healthcare purposes: a survey study from consumers' perspectives. *BMC Medical Informatics and Decision Making*, 20(1), 170. doi: 10.1186/s12911-020-01191-1.

Ettinger, M., & Windhagen, H. (2020). Individual revision arthroplasty of the knee joint. *Der Orthopade*, 49(5), 396–402.

Farmer, Z. L., Dominguez, R. J., Mancinelli, C., et al. (2020). Urogynecological surgical mesh implants: New trends in materials, manufacturing, and therapeutic approaches. *International Journal of Pharmaceutics*, 585, 119512.

Feng, Z. H., Li, X. B., Phan, K., et al. (2018). Design of a 3D navigation template to guide the screw trajectory in the spine: A step-by-step approach using Mimics and 3-Matic software. *Journal of Spine Surgery*, 4(3), 645–653.

Fernandez, A. (2020). Artificial Intelligence teaches drugs to target proteins by tackling the induced folding problem. *Molecular Pharmaceutics*, 17(8), 2761–2767.

Fertleman, C., Aubugeau, W. P., Sher, C., et al. (2018). A discussion of virtual reality as a new tool for training healthcare professionals. *Frontiers in Public Health*, 6, 44.

Fujioka, T., Mori, M., Kubota, K., et al. (2019). Breast ultrasound image synthesis using deep convolutional generative adversarial networks. *Diagnostics*, 9(4), 176.

Gibby, J., Cvetko, S., Javan, R., et al. (2020). Use of augmented reality for image-guided spine procedures. *European Spine Journal*, 29(8), 1823–1832.

Gomez, J. M., Estades, F. J., Meschian, C. S., et al. (2019). Internal hemipelvectomy and reconstruction assisted by 3D printing technology using premade intraoperative cutting and placement guides in a patient with pelvic sarcoma: A case report. *JBJS Case Connect*, 9(4), e60.

Goo, H. W., Park, S. J., & Yoo, S. J. (2020). Advanced medical use of three-dimensional imaging in congenital heart disease: Augmented reality, mixed reality, virtual reality, and three-dimensional printing. *Korean Journal of Radiology*, 21(2), 133–145.

Gulati, S., Emmanuel, A., Patel, M., et al. (2020). Artificial intelligence in luminal endoscopy. *Therapeutic Advances in Gastrointestinal Endoscopy*, 13, 2631774520935220.

Hart, S. N., Flotte, W., Norgan, A. P., et al. (2019). Classification of melanocytic lesions in selected and whole-slide images via convolutional neural networks. *Journal of Pathology Informatics*, 10, 5.

Hashimoto, D. A., Witkowski, E., Gao, L., et al. (2020). Artificial intelligence in anesthesiology: Current techniques, clinical applications, and limitations. *Anesthesiology*, 132(2), 379–394.

He, Y. S., Su, J. R., Li, Z., et al. (2019). Application of artificial intelligence in gastrointestinal endoscopy. *Journal of Digestive Diseases*, 20(12), 623–630.

Held, J., Yu, K., Pyles, C., et al. (2020). Augmented reality-based rehabilitation of gait impairments: Case report. *JMIR mHealth and uHealth*, 8(5), e17804.

Heydon, P., Egan, C., Bolter, L., et al. (2021). Prospective evaluation of an artificial intelligence-enabled algorithm for automated diabetic retinopathy screening of 30,000 patients. *British Journal of Ophthalmology*, 105(5), 723–728.

Houssami, N., Turner, R. M., & Morrow, M. (2017). Meta-analysis of pre-operative magnetic resonance imaging (MRI) and surgical treatment for breast cancer. *Breast Cancer Research and Treatment*, 165(2), 273–283.

Hu, H. Z., Feng, X. B., Shao, Z. W., et al. (2019). Application and prospect of mixed reality technology in the medical field. *Current Medical Science*, 39(1), 1–6.

Hwang, Y., Lee, H. H., Park, C., et al. (2020). An improved classification and localization approach to small bowel capsule endoscopy using convolutional neural network. *Digestive Endoscopy*, 33(4), 598–607.

Iizuka, O., Kanavati, F., Kato, K., et al. (2020). Deep learning models for histopathological classification of gastric and colonic epithelial tumors. *Scientific Reports*, 10(1), 1504.

Jakopec, M., Harris, S. J., Rodriguez, Y. B., et al. (2001). The first clinical application of a "hands-on" robotic knee surgery system. *Computer Aided Surgery*, 6(6), 329–339.

Jiang, L., Wu, Z., Xu, X., et al. (2021). Opportunities and challenges of artificial intelligence in the medical field: Current application, emerging problems, and problem-solving strategies. *Journal of International Medical Research*, 49(3). https://doi.org/10.1177/030006 05211000157

Jiang, Y. Q., Xiong, J. H., Li, H. Y., et al. (2020). Recognizing basal cell carcinoma on smartphone-captured digital histopathology images with a deep neural network. *British Journal of Dermatology*, 182(3), 754–762.

Jones, R., Dobbs, R. W., Halgrimson, W. R., et al. (2020). Single port robotic radical prostatectomy with the da Vinci SP platform: A step-by-step approach. *Canadian Journal of Urology*, 27(3), 10263–10269.

Kamdar, N., & Jalilian, L. (2020). Telemedicine: A digital interface for perioperative anesthetic care. *Anesthesia and Analgesia*, 130(2), 272–275.

Kanavati, F., Toyokawa, G., Momosaki, S., et al. (2020). Weakly-supervised learning for lung carcinoma classification using deep learning. *Scientific Reports*, 10(1), 9297.

Kashyap, A., Kadur, S., Mishra, A., et al. (2018). Cervical pedicle screw guiding jig, an innovative solution. *Journal of Clinical Orthopaedics and Trauma*, 9(3), 226–229.

Kaul, V., Enslin, S., & Gross, S. A. (2020). The history of artificial intelligence in medicine. *Gastrointestinal Endoscopy*, 92(4), 807–812. doi:10.1016/j.gie.2020.06.040.

Ke, Y. Y., Peng, T. T., Yeh, T. K., et al. (2020). Artificial intelligence approach fighting COVID-19 with repurposing drugs. *Biomedical Journal*, 43(4), 355–362.

Khanna, S., Sattar, A., & Hansen, D. (2013). Artificial intelligence in health-the three big challenges. *Australasian Medical Journal*, 6(5), 315.

Kim, J., Zhang, J., Cha, Y., et al. (2020). Advanced bioinformatics rapidly identifies existing therapeutics for patients with coronavirus disease-2019 (COVID-19). *Journal of Translational Medicine*, 18(1), 257.

Komura, D., & Ishikawa, S. (2019). Machine learning approaches for pathologic diagnosis. *Virchows Archiv*, 475(2), 131–138.

Kosaraju, S. C., Hao, J., Koh, H. M., et al. (2020). Deep-Hipo: Multi-scale receptive field deep learning for histopathological image analysis. *Methods*, 179, 3–13.

Laï, M. C., Brian, M., & Mamzer, M. F. (2020). Perceptions of artificial intelligence in healthcare: Findings from a qualitative survey study among actors in France. *Journal of Translational Medicine*, 18(1), 1–13.

Laptev, V. A., Ershova, I. V., & Feyzrakhmanova, D. R. (2022). Medical applications of artificial intelligence (legal aspects and future prospects). *Laws*, 11(1), 3. https://doi.org/10.3390/laws11010003

Lee, S. I., Celik, S., Logsdon, B. A., Lundberg, S. M., Martins, T. J., Oehler, V. G., Estey, E. H., Miller, C. P., Chien, S., Dai, J., Saxena, A., Blau, C. A., & Becker, P. S. (2018). Machine learning approach to integrate big data for precision medicine in acute myeloid leukemia. *Nature Communications*, 9(1), 42.

Lenfant, L., Wilson, C. A., Sawczyn, G., et al. (2020). Single-port robot-assisted dismembered pyeloplasty with mini-Pfannenstiel or peri-umbilical access: Initial experience in a single center. *Urology*, 143, 147–152.

Levin, D., Mackensen, G. B., Reisman, M., et al. (2020). 3D printing applications for transcatheter aortic valve replacement. *Current Cardiology Reports*, 22(4), 23.

Li, X., Yao, R., Tan, X., et al. (2019). Molecular and phenotypic spectrum of Noonan syndrome in Chinese patients. *Clinical Genetics*, 96(4), 290–299.

Liang, G., Fan, W., Luo, H., et al. (2020). The emerging roles of artificial intelligence in cancer drug development and precision therapy. *Biomedicine & Pharmacotherapy*, 128, 110255.

Liu, P. R., Lu, L., Zhang, J. Y., Huo, T. T., Liu, S. X., & Ye, Z. W. (2021). Application of artificial intelligence in medicine: An overview. *Current Medical Science*, 41(6), 1105–1115. doi:10.1007/s11596-021-2474-3. PMID: 34874486; PMCID: PMC8648557.

López-Robles, J. R., Otegi-Olaso, J. R., Gómez, I. P., & Cobo, M. J. (2019). 30 years of intelligence models in management and business: A bibliometric review. *International Journal of Information Management*, 48, 22–38.

Mali, S. N., Pratap, A. P. (2021). Targeting infectious coronavirus disease 2019 (COVID-19) with Artificial Intelligence (AI) applications: Evidence based opinion. *Infectious Disorders – Drug Targets*, 21(4), 475–477. doi: 10.2174/1871526520666200622144857. PMID: 32568026.

Manne, R., & Kantheti, S. (2021). Application of artificial intelligence in healthcare: Chances and challenges. *Current Journal of Applied Science and Technology*, 40, 78–89. doi:10.9734/CJAST/2021/v40i631320.

McKinney, S. M., Sieniek, M., Godbole, V., et al. (2020). International evaluation of an AI system for breast cancer screening. *Nature*, 577(7788), 89–94.

Meskó, B., & Görög, M. (2020). A short guide for medical professionals in the era of artificial intelligence. *npj Digital Medicine*, 3(1), 126. https://doi.org/10.1038/s41746-020-00333-z

Mintz, Y., & Brodie, R. (2019). Introduction to artificial intelligence in medicine. *Minimally Invasive Therapy & Allied Technologies*, 28(2), 73–81. doi:10.1080/13645706.2019.1575882.

Mirchi, N., Bissonnette, V., Ledwos, N., et al. (2020). Artificial neural networks to assess virtual reality anterior cervical discectomy performance. *Operative Neurosurgery,* 19(1), 65–7586.

Namikawa, K., Hirasawa, T., Yoshio, T., et al. (2020). Utilizing artificial intelligence in endoscopy: A clinician's guide. *Expert Review of Gastroenterology & Hepatology*, 14(1), 1–18.

Nas, S., & Koyuncu, M. (2019). Emergency department capacity planning: A recurrent neural network and simulation approach. *Computational and Mathematical Methods in Medicine*, 2019, 4359719.

Navarrete, A. J., & Hashimoto, D. A. (2020). Current applications of artificial intelligence for intraoperative decision support in surgery. *Frontiers in Medicine*, 14(4), 369–381.

Nguyen, D. T., Kang, J. K., Pham, T. D., et al. (2020). Ultrasound image-based diagnosis of malignant thyroid nodule using artificial intelligence. *Sensors*, 20(7), 182240.

Nguyen, D. T., Pham, T. D., Batchuluun, G., et al. (2019). Artificial intelligence-based thyroid nodule classification using information from spatial and frequency domains. *Journal of Clinical Medicine*, 8(11), 1976.

Nikoyan, L., & Patel, R. (2020). Intraoral scanner, three-dimensional imaging, and three-dimensional printing in the dental office. *Dental Clinics of North America*, 64(2), 365–378.

Pandey, M., Choudhury, H., Fern, J., et al. (2020). 3D printing for oral drug delivery: a new tool to customize drug delivery. *Drug Delivery and Translational Research*, 10(4), 986–1001.

Park, J. W., Kang, H. G., Kim, J. H., et al. (2021). The application of 3D-printing technology in pelvic bone tumor surgery. *Journal of Orthopaedic Science*, 26(2), 276–283.

Patel, V. L., Shortliffe, E. H., Stefanelli, M., et al. (2009). The coming of age of artificial intelligence in medicine. *Artificial Intelligence in Medicine*, 46(1), 5–17. doi:10.1016/j.artmed.2008.07.017.

Poncette, A. S., Mosch, L., Spies, C., et al. (2020). Improvements in patient monitoring in the intensive care unit: Survey study. *Journal of Medical Internet Research*, 22(6), e19091.

Ramezani, R., Zhang, W., Xie, Z., et al. (2019). A combination of indoor localization and wearable sensor-based physical activity recognition to assess older patients undergoing subacute rehabilitation: Baseline study results. *JMIR mHealth and uHealth*, 7(7), e14090.

Rey, F., Barzaghini, B., Nardini, A., et al. (2020). Advances in tissue engineering and innovative fabrication techniques for 3-D-structures: Translational applications in neurodegenerative diseases. *Cells*, 9(7), 1636.

Rodriguez, R. A., Lang, K., Gubern, M. A., et al. (2019a). Can we reduce the workload of mammographic screening by automatic identification of normal exams with artificial intelligence? A feasibility study. *European Radiology*, 29(9), 4825–4832.

Rodriguez, R. A., Lang, K., Gubern, M. A., et al. (2019b). Stand-alone artificial intelligence for breast cancer detection in mammography: Comparison with 101 radiologists. *Journal of the National Cancer Institute*, 111(9), 916–922.

Rojas, M. E., Cabrera, M. E., Lin, C., et al. (2020). The System for Telementoring with Augmented Reality (STAR): A head-mounted display to improve surgical coaching and confidence in remote areas. *Surgery*, 167(4), 724–731.

Russo, G., Reche, P., Pennisi, M., et al. (2020). The combination of artificial intelligence and systems biology for intelligent vaccine design. *Expert Opinion on Drug Discovery*, 1–15.

Saab, A., Saikali, M., & Lamy, J. B. (2020). Comparison of machine learning algorithms for classifying adverse-event related to 30-day hospital readmissions: Potential implications for patient safety. *Studies in Health Technology and Informatics*, 272, 51–54.

Sadeghi, A. H., Taverne, Y., Bogers, A., et al. (2020). Immersive virtual reality surgical planning of minimally invasive coronary artery bypass for Kawasaki disease. *European Heart Journal*, 41(34), 3279.

Sakagianni, A., Feretzakis, G., Kalles, D., et al. (2020). Setting up an Easy-to-use machine learning pipeline for medical decision support: A case study for COVID-19 diagnosis based on deep learning with CT scans. *Studies in Health Technology and Informatics*, 272, 13–16.

Salah, M., Tayebi, L., Moharamzadeh, K., et al. (2020). Three-dimensional bio-printing and bone tissue engineering: Technical innovations and potential applications in maxillofacial reconstructive surgery. *Maxillofacial Plastic and Reconstructive Surgery*, 42(1), 18.

Salmas, M., Chronopoulos, E., & Chytas, D. (2020). Comment on: "A novel evaluation model for a mixed-reality surgical navigation system: Where Microsoft HoloLens meets the operating room." *Surgical Innovation*, 27(3), 297–299.

Samareh, A., Chang, X., Lober, W. B., et al. (2019). Artificial intelligence methods for surgical site infection: Impacts on detection, monitoring, and decision making. *Surgical Infections*, 20(7), 546–554.

Sappenfield, J. W., Smith, W. B., Cooper, L. A., et al. (2018). Visualization improves supraclavicular access to the subclavian vein in a mixed reality simulator. *Anesthesia & Analgesia*, 127(1), 83–89.

Secinaro, S., Calandra, D., Secinaro, A., et al. (2021). The role of artificial intelligence in healthcare: A structured literature review. *BMC Medical Informatics and Decision Making*, 21(1), 125. https://doi.org/10.1186/s12911-021-01488-9

Seger, C., & Cannesson, M. (2020). Recent advances in the technology of anesthesia. *F1000Research*, 9, F1000 Faculty Rev-375.

Serag, A., Ion, M. A., Qureshi, H., et al. (2019). Translational AI and deep learning in diagnostic pathology. *Frontiers in Medicine*, 6, 185.

Sharma, P., Pante, A., & Gross, S. A. (2020). Artificial intelligence in endoscopy. *Gastrointestinal Endoscopy*, 91(4), 925–931.

Skelley, N. W., Smith, M. J., Ma, R., et al. (2019). Three-dimensional printing technology in orthopaedics. *Journal of the American Academy of Orthopaedic Surgeons*, 27(24), 918–925.

Sorrentino, F. S., Jurman, G., De NK, et al. (2020). Application of artificial intelligence in targeting retinal diseases. *Current Drug Targets*, 21(12), 1208–1215.

Stefanelli, L. V., Mandelaris, G. A., Franchina, A., et al. (2020). Accuracy evaluation of 14 maxillary full arch implant treatments performed with Da Vinci Bridge: A case series. *Materials*, 13(12), 2806.

Stefano, G. B. (2017). Robotic surgery: Fast forward to telemedicine. *Medical Science Monitor*, 23, 1856.

Stoel, B. C. (2019). Artificial intelligence in detecting early RA. *Seminars in Arthritis and Rheumatism*, 49(3S), S25–S28.

Sun, C., Zhang, Y., Chang, Q., et al. (2020). Evaluation of a deep learning-based computer-aided diagnosis system for distinguishing benign from malignant thyroid nodules in ultrasound images. *Medical Physics*, 47(9), 3952–3960.

Sun, R., Limkin, E. J., Vakalopoulou, M., et al. (2018). A radiomics approach to assess tumor-infiltrating CD8 cells and response to anti-PD-1 or anti-PD-L1 immunotherapy: An imaging biomarker, retrospective multicohort study. *Lancet Oncology*, 19(9), 1180–1191.

Tae, K. (2020). Robotic thyroid surgery. *Auris Nasus Larynx*, 48(3), 331–338.

Takakusagi, Y., Takakusagi, K., Sakaguchi, K., et al. (2020). Phage display technology for target determination of small-molecule therapeutics: an update. *Expert Opinion on Drug Discovery*, 1–13.

Tejo, O. A., Buj, C. I., & Fenollosa, A. F. (2020). 3D printing in medicine for preoperative surgical planning: A review. *Annals of Biomedical Engineering*, 48(2), 536–555.

Tomita, K., Nagao, R., Touge, H., et al. (2019). Deep learning facilitates the diagnosis of adult asthma. *Allergology International*, 68(4), 456–461.

Tsintavi, E., Rekkas, D. M., Bettini, R. (2020). Partial tablet coating by 3D printing. *International Journal of Pharmaceutics*, 581, 119298.

Tursunbayeva, A., & Renkema, M. (2022). Artificial intelligence in health-care: Implications for the job design of healthcare professionals. *Asia Pacific Journal of Human Resources*. https://doi.org/10.1111/1744-7941.12325

Vaishya, R., Javaid, M., Khan, I. H., et al. (2020). Artificial Intelligence (AI) applications for COVID-19 pandemic. *Diabetes & Metabolic Syndrome: Clinical Research & Reviews*, 14(4), 337–339.

van de Sande, D., Van Genderen, M. E., Smit, J. M., et al. (2022). Developing, implementing and governing artificial intelligence in medicine: A step-by-step approach to prevent an artificial intelligence winter. *BMJ Health & Care Informatics*, 29, e100495. https://doi.org/10.1136/bmjhci-2021-100495

Vellido, A. (2019). Societal issues concerning the application of artificial intelligence in medicine. *Kidney Disease*, 5, 11–17. https://doi.org/10.1159/000492428

Vidal, L., Kampleitner, C., Brennan, M. A., et al. (2020). Reconstruction of large skeletal defects: Current clinical therapeutic strategies and future directions using 3D printing. *Frontiers in Bioengineering and Biotechnology*, 8, 61.

Wang, C., Zhang, L., Qin, T., et al. (2020). 3D printing in adult cardiovascular surgery and interventions: A systematic review. *Journal of Thoracic Disease*, 12(6), 3227–3237.

Wang, S., Yang, D. M., Rong, R., et al. (2019). Artificial intelligence in lung cancer pathology image analysis. *Cancers*, 11(11), 1673.

Wu, D., Xiang, Y., Wu, X., et al. (2020). Artificial intelligence-tutoring problem-based learning in ophthalmology clerkship. *Annals of Translational Medicine*, 8(11), 700.

Wu, X., Liu, R., Yu, J., et al. (2018). Mixed reality technology launches in orthopedic surgery for comprehensive preoperative management of complicated cervical fractures. *Surgical Innovation*, 25(4), 421–422.

Xie, Q., Liu, Y., Huang, H., et al. (2020). An innovative method for screening and evaluating the degree of diabetic retinopathy and drug treatment based on artificial intelligence algorithms. *Pharmacological Research*, 159, 104986.

Xing, F., Xiang, Z., Rommens, P. M., et al. (2020). 3D bioprinting for vascularized tissue-engineered bone fabrication. *Materials*, 13(10), 2278.

Yamaguchi, J. T., & Hsu, W. K. (2019). Three-dimensional printing in minimally invasive spine surgery. *Current Reviews in Musculoskeletal Medicine*, 12(4), 425–435.

Yang, Y. Y., & Shulruf, B. (2019). Expert-led and artificial intelligence (AI) system-assisted tutoring course increase confidence of Chinese medical interns on suturing and ligature skills: Prospective pilot study. *Journal of Educational Evaluation for Health Professions*, 16, 7.

Yoshida, S., Sugimoto, M., Fukuda, S., et al. (2020). Mixed reality computed tomography-based surgical planning for partial nephrectomy using a head-mounted holographic computer. *International Journal of Urology*, 26(6), 681–682.

Zeiger, J., Costa, A., Bederson, J., et al. (2020). Use of mixed reality visualization in endoscopic endonasal skull base surgery. *Operative Neurosurgery (Hagerstown)*, 19(1), 43–52.

Zhang, H. T., Zhang, J. S., Zhang, H. H., et al. (2020). Automated detection and quantification of COVID-19 pneumonia: CT imaging analysis by a deep learning-based software. *European Journal of Nuclear Medicine and Molecular Imaging*, 47(11), 2525–2532.

Zhao, Y., Liang, C., Gu, Z., et al. (2020). A new design scheme for intelligent upper limb rehabilitation training robot. *International Journal of Environmental Research and Public Health*, 17(8), 2948.

Zemouri, R., Devalland, C., Valmary, D. S., et al. (2019). Neural network: A future in pathology? *Annals of Pathology*, 39(2), 119–129.

Zhavoronkov, A. (2020). Medicinal chemists versus machines challenge: What will it take to adopt and advance artificial intelligence for drug discovery? *Journal of Chemical Information and Modeling*, 60(6), 2657–2659.

Zhou, F., Xue, F., & Zhang, S. (2020). The application of 3D printing patient specific instrumentation model in total knee arthroplasty. *Saudi Journal of Biological Sciences*, 27(5), 1217–1221.

Zuo, S., & Yang, G. Z. (2017). Endomicroscopy for computer and robot-assisted intervention. *IEEE Reviews in Biomedical Engineering*, 10, 12–25.

2 Open Problems of XAI Especially for Medical Domain

Ferdi Sarac

2.1 INTRODUCTION

Recently, artificial intelligence (AI) has shown great promise in medicine (Zhang et al., 2022). On the other hand, the use of AI applications, especially in clinical applications, becomes difficult due to explainability problems.

Explainable artificial intelligence (XAI) concept has emerged to overcome interpretability and transparency problems in AI applications. Recently, XAI has attracted the attention of researchers; specifically, XAI studies in medicine have dramatically increased. To show this, we carried out a literature review on Pubmed from 2018 to 2022 using the keywords "Explainable Artificial Intelligence". The results are shown in Figure 2.1. Despite the recent popularity of XAI, there are challenges in achieving explainability, especially in biomedical data science. In this chapter, therefore, we present the recent open problems in XAI. We revealed these problems by examining the literature and made some contributions. Finally, we provide a taxonomy of recent problems of XAI to assist future practitioners in their research.

This chapter is organized as follows. Section 2 describes fundamental terms of XAI, Section 3 presents recent limitations of problems of XAI methods and a taxonomy of recent problems of XAI, and Section 4 concludes the chapter.

2.2 FUNDAMENTAL DEFINITIONS

In this section, Machine Learning, AI, XAI and essential terms of XAI are briefly defined.

AI: AI can be defined as computer-assisted systems that simulate human intelligence and perform thought-required tasks that humans can do.

Machine Learning: Machine learning can be described as a branch of AI that allows computer systems to learn from experience without being explicitly programmed.

XAI: While there is no unified accepted technical definition of XAI, it can simply be described as explanation methods that attempt to create AI systems which are more perceptible to humans.

DOI: 10.1201/9781003426073-2

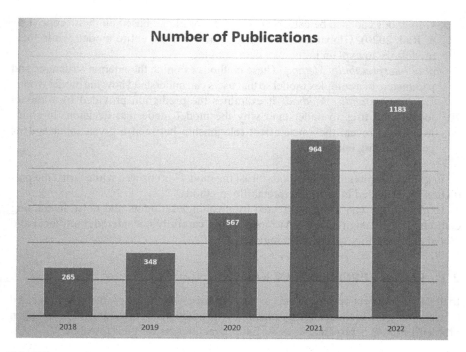

FIGURE 2.1 The number of XAI studies published on Pubmed from 2018 to 2022.

Explainability: Explainability decodes the internal mechanics of a machine learning model so that it becomes more transparent to the users. This decoding can be considered as a transition key that converts the machine learning model from a black box to a white box. Explainability in AI consists of transparency and trust (de Bruijn, et al., 2022). Trust can be affected by situational factors and transparency which may either increase or decrease trust. Furthermore, XAI itself may increase or decrease trust (Bannister & Connolly, 2011).

Interpretability: Explainability and Interpretability are two main terms that distinguish XAI from AI. These terms are generally thought to have the same meaning (Islam et al., 2021; Linardatos et al., 2020). They are also often used interchangeably (Hansen & Rieger, 2019; Miller, 2019). In this study, we do the same.

Explainability methods can be divided into two categories: intrinsic and post hoc.

Intrinsic Methods: These models incorporate interpretability directly into their internal mechanisms. Intrinsic explainability models are generated by designing self-explanatory models (Chaddad et al., 2023). Therefore, these methods are specific to architectural models.

Post-Hoc Methods: In post-hoc methods, a second model is needed for explanations. The post-hoc global explanation aims to provide a general understanding of what information is acquired by pre-trained models. Since post-hoc models exploit a second model for explanation, they ensure that prediction performance is maintained (Madsen et al., 2022).

XAI methods can also be categorized as local or global based on their scope (Das & Rad, 2020). Global methods try to understand the entire model, while local models try to explain individual data samples.

Global Interpretability Method: These methods examine the internal structures and parameters of a complex model so that users can understand how the model works.

Local Interpretability Method: It examines the prediction provided by a model locally and tries to understand why the model made that decision. In short, it helps to reveal the cause-effect relationship between a given input and its corresponding output.

In global interpretability methods "trust the model" is ensured while "trust the prediction" is provided in local interpretability methods.

Explainability can also be affected by meaningless inputs and noise in training datasets. If the design of the XAI model is not carefully considered, both local and global explainability can be meaningless.

2.3   RECENT PROBLEMS IN XAI

In this section recent open problems in XAI are presented. These problems are revealed by examining the literature and by making some additions. A taxonomy of recent open problems in XAI is provided to assist future practitioners for their research.

2.3.1   METHOD DESIGN

Method design poses a particular problem for post-hoc explainability. The explanations should express the mechanism of the AI model, that is, the real working situation. However, post-hoc methods approximate the behavior of machine learning models in order to provide explainability. This approximation can be insufficient to fully describe the state of the machine learning model. Furthermore, explanation may even be unrelated to the model.

2.3.2   INTERPRETABILITY EVALUATION

One of the main problems of internal explanation is the evaluation of interpretability. There are different metrics to measure the interpretability of XAI methods. Each metric has its own pros and cons. However, there is no consensus on when to use which metrics; in fact, selection of metrics is rather related to the task.

Unfortunately, there is no real consensus on what interpretability is in machine learning, and it is not clear how to measure it. However, there is an attempt to provide evaluation metrics to measure interpretability. Doshi-Velez and Kim (2017) proposed three different evaluation approaches: Application-grounded evaluation, Human-grounded evaluation and Functionality-grounded evaluation.

Application-Grounded Evaluation: In this type of evaluation, end-users conduct the assessment with a real application task, and the performance of XAI model is inspected by humans. This evaluation is the most appropriate form of assessment as it receives support from experts for evaluation of interpretability. However, it is very costly to apply and often not feasible. Furthermore, comparison of results in different domains is difficult (Carvalho et al., 2019).

Human-Grounded Evaluation: This evaluation is similar to application-grounded evaluation, however, unlike application-grounded evaluation, experiments are conducted with laypersons, not domain experts. These assessments are less expensive than application-grounded evaluations.

Functionality-Grounded Evaluation: In this type of evaluation, which does not require human experimentation, some formal definitions of interpretability are used as proxy to assess the quality of explanations. This evaluation method is less costly than both application-grounded evaluation and human-grounded evaluation methods. It is especially suitable to be used for immature methods or in situations where the use of human subjects is unethical.

In both human-grounded and application-grounded evaluations, experiments are carried out with either end-users or lay humans, in order to evaluate the quality of explanations. Therefore, subjective measures, such as confidence and trust are mainly emphasized points to assess explainability of XAI models (Ribeiro et al., 2016; Weitz et al., 2019; Zhou et al., 2019, 2021).

We provided a novel taxonomy of evaluation approaches. We classified these approaches based on four different parameters: Human Intervention, which indicates whether human experiments are utilized for evaluations; Complexity of Tasks presents how difficult it is to perform evaluations; Cost shows cost of (time and budget) applying evaluations; and Reliability implies the reliability levels of different evaluation approaches. The taxonomy is presented in Table 2.1.

XAI methods can sometimes produce unexpected explanations. There are usually two main reasons for this. The first is due to the unreliable behaviors of the model. The second is thanks to the inadequacy of the explanation method. Therefore, as mentioned before, there is a need to generate innovative metrics that can better determine the accuracy of explanations.

2.3.3 LACK OF EVALUATION

In the previous subsection, we presented three approaches used for the evaluation of XAI models. Each of these approaches has advantages and disadvantages. Since

TABLE 2.1
A Taxonomy of XAI Evaluation Approaches

Evaluation Level	Human Intervention	Complexity of Tasks	Cost	Reliability
Application-Grounded	Yes	High	High	High
Human-Grounded	Yes	Medium	Medium	Medium
Functionality-Grounded	No	Low	Low	Low

TABLE 2.2
Some Recent XAI Studies in Medicine

Author/s	Objective	XAI Method	Evaluation Method	Year
(Yagin et al., 2023)	Identification of discriminative metabolites for Myalgic encephalomyelitis/chronic fatigue syndrome	SHAP	N/A	2023
(Kumar & Das 2023)	identify Peripheral blood mononuclear cell derived biomarkers of Breast cancer	SHAP	N/A	2023
Raihan & Nahid, 2022).	Malaria cell image classification	SHAP	N/A	2022
(Sarp et al., 2021)	Chronic wound	LIME	N/A	2021
(El-Sappagh et al., 2021)	Alzheimer's disease	SHAP, Fuzzy	N/A	2021
(Chen et al., 2020)	Clinical diagnosis	Bayesian network ensembles	Application-Grounded	2020
(Sabol et al., 2020)	Colorectal cancer diagnosis	Explainable Cumulative Fuzzy Class Membership Criterion (X-CFCMC)	Human-Grounded	2020
(Wei et al., 2020)	Diagnosis of thyroid nodules	CAM	N/A	2020

evaluation of XAI models depends on human cognition, only qualitative evaluation of these models is possible. Unfortunately, most of the recent studies in medicine provide XAI models without evaluation. There have been only a few recent studies where XAI evaluations by domain experts (e.g., radiologists) are provided. In addition, most of the studies conducted in the field of XAI consist of applying existing machine learning methods. Since these methods are performed without the participation of medical specialists, they may not meet the actual needs of physicians. A short list of recent XAI studies in medical domain is presented in Table 2.2.

2.3.4 TRADE-OFF BETWEEN MODEL INTERPRETABILITY AND PERFORMANCE

In biomedical data science applications, it is difficult to achieve high performance. For this reason, explainability is generally not seen as the first priority, especially in biomedical applications. Furthermore, performances of AI models may be mediocre, albeit their explainability is quite good. Interpretability and performance in machine learning models are conflicting concepts. Making a machine learning model more interpretable may ultimately decrease its yielded prediction performance (Coeckelbergh, 2020; Došilović

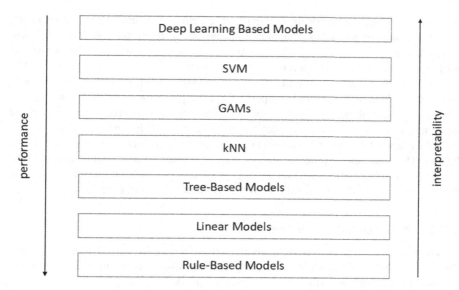

FIGURE 2.2 The relation between interpretability and performance in AI models.

et al., 2018; Molnar, 2021; Wei et al., 2020). Therefore, it is necessary to develop a model that will effectively provide the optimum balance between them (Arrieta et al., 2020). Figure 2.2 illustrates the trade-off between interpretability and performance.

2.3.5 ETHICAL PROBLEMS

Transparency and explainability are extremely important concepts, both ethically (Muller et al., 2021; Stahl & Coeckelbergh, 2016) and legally (Schneeberger et al., 2020). While explainability may enhance understanding and confidence in a system (Adadi & Berrada, 2018; Lipton, 2018; Miller et al., 2017), simple explanations can disguise unwanted features of the system and lead users to draw unethical conclusions (Antoniadi et al., 2021).

If the explanations provided by post-hoc methods are optimized to satisfy subjective needs, these explanations can be potentially misleading. For example, an algorithm may be intentionally or unintentionally optimized to provide misleading yet acceptable explanations. It is quite possible to encounter such behaviors in daily life. For example, we may encounter such situations when accepting students to universities (Lipton, 2018). Explanations in XAI algorithms should be sensitive to ethical values and free from discrimination.

2.3.6 LACK OF END-USER EXPLANATIONS

The main purpose of explicable machine learning models is that the explanations should meet the demands of the end user. However, one of the most important challenges faced in explainable AI is that the methods are generally applied by machine learning engineers to debug models, rather than end users (Bhatt et al., 2020). The perceived quality of explanations will be different for users with different levels of expertise (Yang et al., 2023). It means a novice end-user and a domain expert will have different levels of perception of explanations. Novice users require simplified explanations for them to understand. On the other hand, simplified explanations may be found lacking in depth by experts. For example, local descriptions are usually obtained by describing the contributions of each feature in a given input, since these descriptions will be useful to developers and researchers. However, it is not suitable for end users who cannot perform statistical analysis of feature importance distribution. These aforementioned problems may lead users to distrust the system. XAI methods should not only focus on enhancing users' understanding, but also increase the transparency and trustworthiness of the AI models (Han & Liu, 2021).

2.3.7 Need for Constant Updates

In XAI systems, explanations are generated based on training data. Therefore, as new data comes in or XAI models evolve, explanations may become useless. In order to eliminate this, constant explanation updates are needed.

2.3.8 Problems in Hybrid Approaches

Hybrid systems utilize multiple methods to enhance both robustness and performance. However, different methods might have intrinsic differences in the structure of different methods. These differences may cause inconsistency in hybrid explanation systems.

In order to provide good explainability in hybrid models, it is necessary to combine the non-homogeneous information obtained from different models and to eliminate the inconsistencies that may arise as a result of this combination.

A list of current XAI problems in the medical domain is provided along with the solutions of the problems and affected XAI model types in Table 2.3.

Figure 2.3 illustrates a taxonomy of recent XAI problems in the medical domain.

2.4 CONCLUSION AND FUTURE DIRECTIONS

In this chapter, we covered the essential XAI terms, presented recent open problems in XAI, and provided some suggestions for the recent problems of XAI. We specifically mentioned the loss of evaluation of XAI methods. We examined the recent explainable AI studies in the field of medicine. We noticed that 75% of the studies we reviewed did not evaluate the XAI methods. How can an unevaluated model be reliable? This question will remain as an open question if researchers continue to propose unevaluated XAI models. Even though evaluation methods are applied to assess quality of XAI methods, there is no universally accepted evaluation metric. Therefore, there is a need to create innovative metrics that can better determine the accuracy of explanations. We mentioned ethical problems in XAI. While explainability is increased in XAI applications, it should not cause ethical problems. Since XAI will continue to grow over the next few years, ethical problems in XAI must be fully eliminated.

TABLE 2.3
Recent XAI Problems in Medicine and Its Solutions

Problems	Solutions	Affected Models
Method Design	The explanations should express the mechanism of the artificial intelligence model	Post Hoc
Interpretability Evaluation	New metrics should be generated to evaluate interpretability of the models.	Both
Lack of Evaluation	Explainability of the XAI methods should be evaluated to confirm their effectiveness.	Both
Trade-off Between Model Interpretability and Accuracy	There is a need for new XAI models that will strike the optimum balance between performance and explainability.	Intrinsic
Ethical Problems	Explanations in XAI algorithms must be sensitive to ethical values and free from discrimination.	Both
Lack of End-User Explanations	Explanations in XAI models should meet the demands of the end user.	Both
Need for Constant Updates	Constant explanation updates are needed.	Both
Problems in Hybrid Approaches	Inconsistencies in hybrid XAI models must be eliminated.	Both

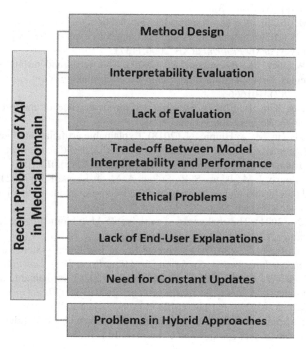

FIGURE 2.3 Recent problems of XAI in medical domain.

REFERENCES

Adadi, A., & Berrada, M. (2018). Peeking inside the black-box: A survey on explainable artificial intelligence (XAI). *IEEE Access*, 6, 52138–52160.

Antoniadi, A. M., Du, Y., Guendouz, Y., Wei, L., Mazo, C., Becker, B. A., & Mooney, C. (2021). Current challenges and future opportunities for XAI in machine learning-based clinical decision support systems: A systematic review. *Applied Sciences*, 11(11), 5088.

Arrieta, A. B., Díaz-Rodríguez, N., Del Ser, J., Bennetot, A., Tabik, S., Barbado, A., ... & Herrera, F. (2020). Explainable Artificial Intelligence (XAI): Concepts, taxonomies, opportunities and challenges toward responsible AI. *Information Fusion*, 58, 82–115.

Bannister, F., & Connolly, R. (2011). Trust and transformational government: A proposed framework for research. *Government Information Quarterly*, 28(2), 137–147.

Bhatt, U., Xiang, A., Sharma, S., Weller, A., Taly, A., Jia, Y., Ghosh, J., Puri, R., Moura, J. M. F., Eckersley, P. (2020). Explainable machine learning in deployment. In: FAT*'20: conference on fairness, accountability, and transparency, Barcelona, Spain, January 27–30, 2020, ACM, pp 648–657.

Carvalho, D. V., Pereira, E. M., & Cardoso, J. S. (2019). Machine learning interpretability: A survey on methods and metrics. *Electronics*, 8(8), 832.

Chaddad, A., Peng, J., Xu, J., & Bouridane, A. (2023). Survey of explainable AI techniques in healthcare. *Sensors*, 23(2), 634.

Chen, J., Dai, X., Yuan, Q., Lu, C., & Huang, H. (2020, July). Towards interpretable clinical diagnosis with Bayesian network ensembles stacked on entity-aware CNNs. In *Proceedings of the 58th Annual Meeting of the Association for Computational Linguistics* (pp. 3143–3153).

Coeckelbergh, M. (2020). *AI ethics*. Cambridge: The MIT Press.

Das, A., & Rad, P. (2020). Opportunities and challenges in explainable artificial intelligence (xai): A survey. *arXiv Preprint* arXiv:2006.11371.

de Bruijn, H., Warnier, M., & Janssen, M. (2022). The perils and pitfalls of explainable AI: Strategies for explaining algorithmic decision-making. *Government Information Quarterly*, 39(2), 101666.

Doshi-Velez, F., & Kim, B. (2017). Towards a rigorous science of interpretable machine learning. *arXiv Preprint* arXiv:1702.08608.

Došilović, F. K., Brčić, M., & Hlupić, N. (2018). Explainable artificial intelligence: A survey. In *2018 41st International Convention on Information and Communication Technology, Electronics and Microelectronics (MIPRO)* (pp. 0210-0215). IEEE.

El-Sappagh, S., Alonso, J. M., Islam, S. R., Sultan, A. M., & Kwak, K. S. (2021). A multilayer multimodal detection and prediction model based on explainable artificial intelligence for Alzheimer's disease. *Scientific Reports*, 11(1), 2660.

Han, H., & Liu, X. (2021). The challenges of explainable AI in biomedical data science. *BMC Bioinformatics*, 22(12), 1–3.

Hansen, L. K., & Rieger, L. (2019). Interpretability in intelligent systems–a new concept? *Explainable AI: Interpreting, Explaining and Visualizing Deep Learning*, 41–49.

Islam, S. R., Eberle, W., Ghafoor, S. K., & Ahmed, M. (2021). Explainable artificial intelligence approaches: A survey. *arXiv Preprint* arXiv:2101.09429.

Kumar, S., & Das, A. (2023). Peripheral blood mononuclear cell derived biomarker detection using eXplainable Artificial Intelligence (XAI) provides better diagnosis of breast cancer. *Computational Biology and Chemistry*, 104, 107867.

Linardatos, P., Papastefanopoulos, V., & Kotsiantis, S. (2020). Explainable AI: A review of machine learning interpretability methods. *Entropy*, 23(1), 18.

Lipton, Z. C. (2018). The mythos of model interpretability: In machine learning, the concept of interpretability is both important and slippery. *Queue*, 16(3), 31–57.

Madsen, A., Reddy, S., & Chandar, S. (2022). Post-hoc interpretability for neural NLP: A survey. *ACM Computing Surveys*, 55(8), 1–42.

Miller, T. (2019). Explanation in artificial intelligence: Insights from the social sciences. *Artificial Intelligence*, 267, 138.

Miller, T., Howe, P., Sonenberg, L., & AI, E. (2017). Beware of inmates running the asylum or: how i learnt to stop worrying and love the social and behavioural sciences. CoRR abs/1712.00547. arXiv Preprint arXiv:1712.00547.

Molnar, C,. (2019). Interpretable machine learning. [Online]. Available: https://christophm.git hub.io/interpretable-ml-book/

Muller, H., Mayrhofer, M. T., Van Veen, E. B., & Holzinger, A. (2021). The ten commandments of ethical medical AI. *Computer*, 54(07), 119–123.

Raihan, M. J., & Nahid, A. A. (2022). Malaria cell image classification by explainable artificial intelligence. *Health and Technology*, 12(1), 47–58.

Ribeiro, M. T., Singh, S., & Guestrin, C. (2016, August). *"Why should i trust you?" Explaining the predictions of any classifier*. In Proceedings of the 22nd ACM SIGKDD International Conference on Knowledge Discovery and Data Mining (pp. 1135–1144), San Francisco, CA, USA, August 13–17, 2016.

Sabol, P., Sinčák, P., Hartono, P., Kočan, P., Benetinová, Z., Blichárová, A., ... & Jašková, A. (2020). Explainable classifier for improving the accountability in decision-making for colorectal cancer diagnosis from histopathological images. *Journal of Biomedical Informatics*, 109, 103523.

Sarp, S., Kuzlu, M., Wilson, E., Cali, U., & Guler, O. (2021). The enlightening role of explainable artificial intelligence in chronic wound classification. *Electronics*, 10(12), 1406.

Schneeberger, D., Stöger, K., & Holzinger, A. (2020, August). *The European legal framework for medical AI*. In International Cross-Domain Conference for Machine Learning and Knowledge Extraction (pp. 209–226). Cham: Springer International Publishing.

Stahl, B. C., & Coeckelbergh, M. (2016). Ethics of healthcare robotics: Towards responsible research and innovation. *Robotics and Autonomous Systems*, 86, 152–161.

Wei, X., Zhu, J., Zhang, H., Gao, H., Yu, R., Liu, Z., ... & Zhang, S. (2020). Visual interpretability in computer-assisted diagnosis of thyroid nodules using ultrasound images. *Medical Science Monitor: International Medical Journal of Experimental and Clinical Research*, 26, e927007–1.

Weitz, K., Schiller, D., Schlagowski, R., Huber, T., & André, E. (2019, July). *"Do you trust me?" Increasing user-trust by integrating virtual agents in explainable AI interaction design*. In Proceedings of the 19th ACM International Conference on Intelligent Virtual Agents in Paris (pp. 7–9).

Yagin, F. H., Alkhateeb, A., Raza, A., Samee, N. A., Mahmoud, N. F., Colak, C., Yagin, B. (2023). An explainable artificial intelligence model proposed for the prediction of myalgic encephalomyelitis/chronic fatigue syndrome and the identification of distinctive metabolites. Diagnostics, 13(23):3495.

Yang, W., Wei, Y., Wei, H., Chen, Y., Huang, G., Li, X., ... & Kang, B. (2023). Survey on explainable AI: From approaches, limitations and applications aspects. *Human-Centric Intelligent Systems*, 1–28.

Zhang, Y., Weng, Y., & Lund, J. (2022). Applications of explainable artificial intelligence in diagnosis and surgery. *Diagnostics*, 12(2), 237.

Zhou, J., Gandomi, A. H., Chen, F., & Holzinger, A. (2021). Evaluating the quality of machine learning explanations: A survey on methods and metrics. *Electronics*, 10(5), 593.

Zhou, J., Li, Z., Hu, H., Yu, K., Chen, F., Li, Z., & Wang, Y. (2019, May). *Effects of influence on user trust in predictive decision making*. In Extended Abstracts of the 2019 CHI Conference on Human Factors in Computing Systems in Glasgow (pp. 1–6).

3 Explainable AI in Biomedical Applications
Vision, Framework, Anxieties, and Challenges

Azra Nazir, Faisal Rasheed Lone,
Roohie Naaz Mir, and Shaima Qureshi

3.1 MOTIVATION TO EXPLAINABLE AI (XAI)

The collaboration between artificial intelligence (AI) and human experts is envisioned as a partnership to augment healthcare professionals' capabilities and improve overall healthcare delivery (Das & Rad, 2020). However, like every other partnership, it also requires trust as its foundation stone which, unfortunately, is not embedded in typical AI frameworks. Deep learning (DL) algorithms have shown promising results in the healthcare and biomedical field, especially in classifying medical images such as X-rays and CT scans. However, they are still far from being accepted and employed by a common man or medical professional. Several factors contribute to this gap:

Complexity of Diagnosis: Medical diagnosis is a complex task that involves not only the analysis of images but also the integration of patient history, symptoms, and other clinical information. DL algorithms excel at pattern recognition in images but may lack the broader context necessary for accurate diagnosis (Angelov et al., 2021 and Tjoa and Guan, 2020). Radiologists and clinicians bring their expertise, clinical judgment, and knowledge of patient history to make well-informed decisions.

Uncertainty and Ambiguity: Medical imaging often presents cases that are challenging, with ambiguous or subtle abnormalities. DL algorithms can struggle with such cases as they heavily rely on patterns and features learned from training data. Radiologists, on the other hand, possess the ability to interpret nuanced findings, consider differential diagnoses, and evaluate the clinical context to arrive at accurate conclusions.

Limited Generalization: DL algorithms typically generalize well within the range of data they were trained on. However, they may struggle with detecting rare or novel conditions, as they might not have encountered such cases during training. Radiologists, through their experience and exposure to diverse cases, can identify and interpret uncommon abnormalities or variations.

DOI: 10.1201/9781003426073-3

Human Interaction and Communication: Radiologists and clinicians play essential roles in patient care beyond the interpretation of images. They interact directly with patients, understand their concerns, and communicate diagnoses and treatment plans effectively. The empathetic and interpersonal skills of healthcare professionals are crucial for building trust, providing emotional support, and ensuring comprehensive patient care.

Regulatory and Legal Considerations: The deployment of AI algorithms in clinical settings requires rigorous validation, regulatory compliance, and ethical considerations. Regulatory bodies and healthcare institutions have specific guidelines and standards that need to be met to ensure patient safety, data privacy, and liability. The integration of DL algorithms into automated systems involves addressing these requirements, which can be complex and time-consuming.

3.2 UNDERSTANDING XAI

To address these issues, Explainable AI (XAI) has emerged as a critical area of research, aiming to provide understandable and interpretable explanations for AI-driven decision-making processes, as shown in Figure 3.1.

XAI has the potential to bridge the gap of trust between AI algorithms and human experts in the field of medical imaging and beyond. In recent years, the application of AI in the biomedical field has witnessed remarkable advancements (Adadi & Berrada, 2018). However, the inherent complexity and lack of transparency in AI algorithms have raised concerns about their interpretability and reliability. In this

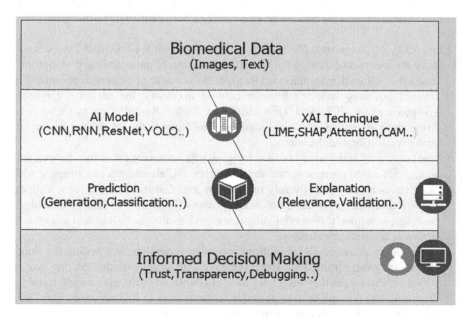

FIGURE 3.1   Workflow of an Explainable AI system in the context of biomedical applications.

chapter, we explore the relevance of XAI in biomedical applications, its advantages, frameworks, limitations, and anxieties.

In the traditional approach, an AI algorithm is trained using a deep-learning neural network to detect lung nodules in chest X-ray images. The AI model learns complex patterns and features from a large dataset of labeled images and develops the ability to identify potential nodules in unseen images (Arrieta et al., 2020). When presented with a new chest X-ray, the AI model produces a prediction indicating the presence or absence of lung nodules. In an XAI approach, the same AI algorithm is employed, but it is enhanced with explainability techniques. Along with providing the prediction, the XAI system generates a visual explanation highlighting the areas of the image that influenced the AI's decision, as shown in Figure 3.2. The explanation could be in the form of a heatmap, indicating regions of higher importance or saliency in the image. This helps the radiologist or clinician understand why the AI model made a particular prediction and which areas of the image contributed to it. The visual explanation enables the radiologist to validate and trust the AI system's decision, ultimately leading to more confident diagnoses. XAI aims to bridge the gap between the complex inner workings of AI algorithms and the need for human understanding and trust, making it particularly valuable in critical domains such as healthcare. XAI can be applied to both deep learning and machine learning algorithms to provide explanations for their predictions. Machine learning algorithms typically rely on statistical models and algorithms to learn patterns and make predictions from data. These algorithms are often based on mathematical functions and can involve techniques such as decision trees, random forests, support vector machines (SVM),

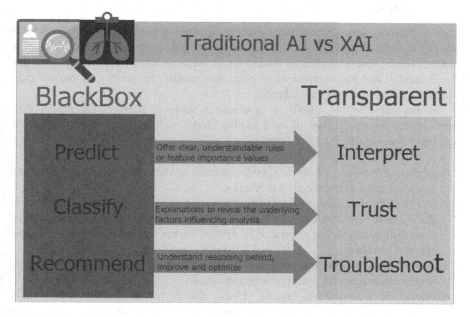

FIGURE 3.2 Transition to explainable AI (XAI).

or logistic regression (Vilone & Longo, 2021). Machine learning models generally have a more interpretable structure, and their decision-making process can be explicitly represented. Deep learning models, such as convolutional neural networks (CNNs) or recurrent neural networks (RNNs), excel at automatically learning complex patterns and features directly from raw data, often achieving state-of-the-art performance (Jiménez-Luna et al., 2020). Deep learning models can have millions of parameters and operate in a more black-box manner, making it challenging to understand their decision-making process. Inherently explainable models, such as decision trees or rule-based systems, offer built-in interpretability and transparency, making them suitable for XAI (Jiménez-Luna et al., 2020). In XAI, the focus is not on modifying existing machine learning algorithms but rather on developing techniques and approaches that provide interpretability and transparency to the decisions made by these algorithms. XAI techniques can be broadly categorized into two types model agnostic approaches and model specific, as discussed next.

3.2.1 MODEL-AGNOSTIC APPROACHES

These approaches can be applied to any AI algorithm without requiring specific modifications. Model-agnostic techniques focus on generating explanations for the predictions made by an AI model without relying on its internal workings. Examples of model-agnostic approaches include permutation importance, partial dependence plots, LIME (Local Interpretable Model-Agnostic Explanations), and SHAP (SHapley Additive exPlanations).

3.2.1.1 LIME (LOCAL INTERPRETABLE MODEL-AGNOSTIC EXPLANATIONS)

LIME is a model-agnostic technique used for explaining the predictions of machine learning models. It provides local explanations by approximating the behavior of a complex model around a specific data point or instance. LIME aims to enhance the interpretability and trustworthiness of black-box models by generating understandable explanations at the instance level.

The key idea behind LIME is to create a simpler, interpretable model that approximates the predictions of the complex model within a local region. Here's a high-level overview of the LIME process:

Selecting an Instance: Choose a specific instance or data point for which you want an explanation of the model's prediction.

Perturbing the Instance: Create perturbations of the selected instance by introducing small random changes while keeping the rest of the data fixed. These perturbed instances are generated by modifying the features or attributes of the original instance.

Model Prediction: Pass the perturbed instances through the complex model and obtain predictions for each perturbed instance.

Creating an Interpretable Model: Using the perturbed instances and their corresponding predictions, fit an interpretable model, such as a linear regression model or decision tree, to approximate the behavior of the complex model in the local region surrounding the selected instance.

Interpreting the Model: Analyze the coefficients or feature importance values of the interpretable model to understand the relative importance of different features in the local prediction. These coefficients provide insights into which features contributed positively or negatively to the prediction, enabling interpretation and understanding.

The explanations generated by LIME are local to a specific instance and provide insights into the model's decision-making process for that instance. LIME does not explain the entire model comprehensively but rather offers instance-specific explanations, which can be useful in scenarios where individual predictions require interpretability.

LIME has been applied to various domains and models, including image classification, text classification, and healthcare. By generating local explanations, LIME helps build trust, transparency, and interpretability in machine learning models, making them more accessible and usable in critical applications.

3.2.1.2 SHAP (SHapley Additive exPlanations)

SHAP is a method for explaining the predictions of machine learning models. It provides a unified framework for explaining the output of any model by assigning importance values to different features or variables. SHAP values are based on the concept of Shapley values from cooperative game theory and offer a rigorous and theoretically grounded approach to feature attribution.

The key idea behind SHAP is to estimate the contribution of each feature to the prediction by considering different combinations of features. Here's a high-level overview of the SHAP process:

Baseline Reference: Define a baseline reference point, which represents the average or default prediction of the model. This baseline is used as a starting point for measuring the contribution of features.

Feature Subsets: Consider all possible subsets of features, ranging from subsets with no features to subsets including all features. Each subset represents a unique combination of features.

Feature Importance Calculation: For each subset of features, compute the contribution or importance of each feature towards the prediction by comparing the model's output with and without the given feature(s). This calculation quantifies the impact of each feature on the prediction.

Shapley Value Estimation: Apply Shapley value principles from cooperative game theory to distribute the feature importance across individual features. The Shapley values ensure fairness and consistency in attributing contributions to each feature in the context of their interactions with other features.

Interpreting the SHAP Values: The resulting SHAP values represent the individual feature contributions to the prediction. Positive SHAP values indicate features that push the prediction higher, while negative values represent features that push it lower. The magnitude of the SHAP value indicates the strength of the influence.

SHAP values can be visualized using various techniques, such as bar charts, summary plots, or dependence plots. These visualizations provide insights into the relative importance and impact of different features on the model's predictions.

SHAP has gained popularity due to its theoretical foundation, flexibility to handle different types of models, and its ability to provide interpretable explanations for individual predictions. It enables users to understand the driving factors behind model predictions, identify important features, detect feature interactions, and assess the robustness of the model's decisions.

3.2.2 INHERENTLY EXPLAINABLE MODELS

Some AI algorithms are inherently more interpretable than others. These models are designed with transparency and explainability in mind, making their decision-making process more easily understandable. Examples of inherently explainable models include decision trees, rule-based systems, and linear models. These models provide explicit rules or feature importance values that aid in understanding their predictions.

3.3 ADVANTAGES OF XAI IN BIOMEDICAL APPLICATIONS

3.3.1 ENHANCED TRUST AND ACCEPTANCE

One of the primary advantages of XAI in the biomedical domain is the ability to enhance the trust and acceptance of AI systems. By providing interpretable explanations, XAI helps clinicians, researchers, and patients understand the reasoning behind AI-driven decisions, leading to increased confidence in the technology. This trust is vital for the widespread adoption of AI in critical biomedical tasks.

3.3.2 IMPROVED CLINICAL DECISION-MAKING

XAI empowers clinicians with valuable insights into AI-driven clinical decision support systems. By generating understandable explanations for diagnostic or treatment recommendations, XAI enables healthcare professionals to evaluate the AI's reasoning and make informed decisions. This can lead to improved accuracy, reduced errors, and enhanced patient outcomes (Samek et al., 2017).

3.3.3 FACILITATING RESEARCH AND DISCOVERY

In drug discovery and development, XAI plays a crucial role in elucidating the relationships between molecular structures, biological targets, and therapeutic efficacy. By providing interpretable explanations for AI-generated predictions, XAI methods assist researchers in understanding the underlying mechanisms and optimizing drug design. This can expedite the discovery of new drugs and therapeutic interventions (Minh et al., 2022).

3.3.4 EMPOWERING PATIENTS

XAI promotes patient empowerment by providing understandable explanations for personalized treatment recommendations. In precision medicine, XAI methods can help patients comprehend the AI's reasoning behind genetic variations, biomarkers, and treatment choices. This enables patients to actively participate in decision-making processes and fosters personalized healthcare.

3.4 FRAMEWORKS FOR XAI IN BIOMEDICAL APPLICATIONS

Several frameworks and techniques have been proposed to achieve explainability in AI systems. In the biomedical world, the following frameworks have shown promise.

3.4.1 RULE-BASED EXPLANATIONS

Rule-based explanations utilize predefined rules to explain AI predictions. These rules can be defined by domain experts or extracted from the underlying AI model. Rule-based frameworks provide human-understandable explanations, enabling clinicians to verify and validate the decision-making process.

A rule-based explanation for an AI model in the context of diagnosing diabetic retinopathy from retinal images.

AI Model: A deep learning model is trained to classify retinal images as either showing signs of diabetic retinopathy or being healthy.

Rule-based Explanation: A rule-based explanation for the AI model's decision could involve a set of predefined rules that align with expert knowledge and medical guidelines as shown in Figure 3.3.

Rule 1: If the image contains microaneurysms (small round red dots) and hemorrhages (red spots or blotches), classify it as diabetic retinopathy.

Rule 2: If there are abnormalities such as exudates (yellow or white lesions) and neovascularization (abnormal blood vessels), classify it as diabetic retinopathy.

Rule 3: If the image shows signs of macular edema (swelling of the central part of the retina), classify it as diabetic retinopathy.

Rule 4: If none of the above abnormalities are present, classify it as a healthy retina.

In this rule-based explanation, the AI model uses a series of if-then rules to classify retinal images. These rules are derived from medical knowledge and guidelines that describe the characteristic features of diabetic retinopathy. By following these rules, the AI model can provide interpretable explanations for its decision-making process.

For instance, if an input image contains microaneurysms, hemorrhages, and exudates, it would be classified as diabetic retinopathy based on Rule 1 and Rule

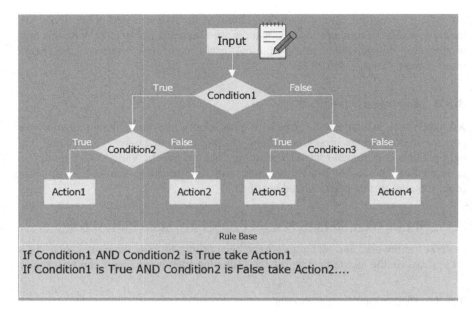

FIGURE 3.3   Illustration of a rule-based system.

2. On the other hand, if the image shows no signs of any of these abnormalities but appears healthy, it would be classified as a healthy retina according to Rule 4. These rule-based explanations provide transparency and interpretability, allowing clinicians to understand why the AI model made a particular diagnosis. By following a set of explicit rules, the AI model's decisions align with medical knowledge, which enhances trust and confidence in its outputs.

3.4.2 FEATURE IMPORTANCE ANALYSIS

Feature importance analysis involves quantifying the contribution of input features to AI predictions. Techniques such as permutation importance, partial dependence plots, and SHAP provide insights into the relevance and impact of different features on the output. Visualizations of feature importance help clinicians interpret and trust AI predictions.

Feature importance analysis in XAI helps researchers and clinicians understand which features have the most influence on the model's output. To illustrate feature importance analysis using SHAP, consider a machine learning model designed to predict the risk of heart disease based on patients. SHAP is applied in an AI model by considering the following aspects.

AI Model: A machine learning model is trained to predict the risk of heart disease based on various patient features such as age, gender, blood pressure, cholesterol levels, and exercise habits. Training is done using a dataset obtained from medical reports.

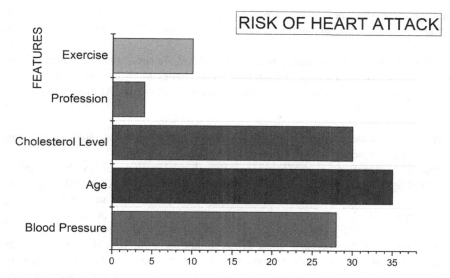

FIGURE 3.4   Illustration of feature importance to predict heart attack.

Feature Importance Analysis with SHAP: SHAP is a technique that provides insights into the contribution of individual features to the AI model's predictions. In this example, SHAP values are calculated for each feature, indicating their impact on the risk of heart disease.

Visualization: The feature importance analysis can be visualized using a bar chart or a summary plot (as shown in Figure 3.4). In this visualization, the length of each bar represents the importance of the corresponding feature. Features with longer bars have a greater influence on the AI model's predictions.

For example, the "Age" feature is shown to have significant importance, as indicated by its long bar. This suggests that age plays a crucial role in determining the risk of heart disease. Similarly, "Cholesterol Levels" and "Blood Pressure" also have notable contributions to the predictions, while "Exercise Habits" has a relatively smaller impact.

By visualizing feature importance using techniques like SHAP, clinicians can interpret the AI model's predictions and understand which patient features are driving the risk assessment. This information helps build trust and confidence in the AI model by providing insights into the relevance and impact of different features on the output. Clinicians can leverage this knowledge to validate the model's predictions and make informed decisions regarding patient care and treatment plans (Deeks, 2019).

3.4.3   VISUAL EXPLANATIONS

Visual explanations utilize visualizations to illustrate the reasoning behind AI predictions. Techniques such as heat maps, saliency maps, or class activation maps highlight the regions or features in medical images that influenced the decision. Visual explanations aid radiologists in understanding and validating AI-based diagnoses or detections (Langer et al., 2021).

Visual explanations using heatmaps are used to illustrate the reasoning behind AI predictions in the context of detecting lung nodules in chest X-ray images. The following steps are required to highlight regions in the X-ray the AI model focuses on when making its prediction.

AI Model: A deep learning model is trained to detect lung nodules, which are potential indicators of lung cancer, in chest X-ray images.

Visual Explanations with Heatmaps: Visual explanations utilize heatmaps to highlight the regions of an image that contributed to the AI model's decision. The heatmaps indicate the areas with higher importance or saliency, providing insights into the features or patterns that influenced the AI's prediction.

In Figure 3.5, the original chest X-ray image is input into the AI model for nodule detection. The heatmap visualization is generated alongside the prediction to highlight the regions in the image that contributed to the AI model's decision. The heatmaps assign higher intensity or color to the areas that played a significant role in the prediction.

The heatmaps can help radiologists and clinicians in the following ways:

Understanding the AI Model's Focus: By examining the heatmap, radiologists can identify the specific regions of interest that the AI model considers important for nodule detection. This information helps radiologists align their attention and focus on the relevant areas during their own interpretation (Nazir et al., 2023 Feb).

Validating AI-based Diagnoses: Heatmaps allow radiologists to verify whether the AI model's predictions align with their own expertise. By comparing their

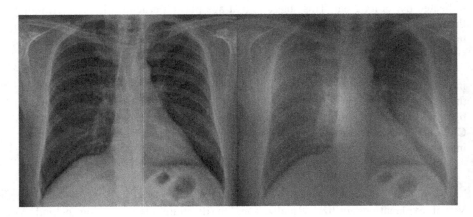

FIGURE 3.5   Visual explanation of an AI-driven diagnosis using heatmap.

own observations with the heatmap, radiologists can assess the consistency and accuracy of the AI model's findings.

Exploring False Positives/Negatives: Heatmaps help in understanding the reasoning behind false positives or false negatives. Radiologists can analyze the heatmap to identify potential reasons for the AI model's incorrect predictions, such as misleading image features or limitations in the training data.

By providing visual explanations with heatmaps, radiologists gain insights into the AI model's decision-making process and can use this information to enhance their own interpretations and validate the AI-based diagnoses or detections (Došilović, Brčić, & Hlupić, 2018, May). This aids in building trust and confidence in the AI system, facilitating its integration into the clinical workflow.

3.5 CASE STUDY: XAI FRAMEWORK FOR INTERPRETATION OF HRCT CHEST ANALYSIS FOR DETECTION OF PULMONARY ARTERIAL HYPERTENSION IN COVID-19 PATIENTS USING CONVOLUTIONAL NEURAL NETWORKS

After COVID-19, there has been a sudden increase in the rate of heart attacks and strokes. The virus can impact the hearts of individuals who were previously healthy. The severe inflammatory response within the body can affect arteries, exacerbating cardiac damage. Research suggests that monitoring the cardiac conditions of individuals who have recovered from COVID-19 is advisable. In our previous study (Nazir et al., 2022, October). VGG16, a popular image classification model, was employed to extract features from the standard axial slice of high-resolution chest CT scans. The CNN model implemented as a black box captures variations in the pulmonary artery region to determine whether a patient is at a high risk of developing pulmonary arterial hypertension or not. A diameter of 27–28 mm is considered normal. In patients with pulmonary arterial hypertension, the diameter has been reported 32–35mm. A deep convolutional neural network can learn this significant deviation to classify patients into high-risk and no-risk classes. The employed model obtained 86.1% accuracy demonstrating the promising potential for further studies analyzing the risks of heart complications in individuals who have recovered from COVID-19. Here we propose the incorporation of Gradient–weighted Class Activation Mapping (Grad-CAM) to identify the regions or features in the high-resolution computed tomography (HRCT) images that influenced the model's decision while classifying the scans. Grad-CAM generates heatmaps that highlight the regions of the image that contribute most significantly to the model's prediction. The advantage of Grad-CAM is that it can be applied to the model without requiring any modifications to the model itself.

By incorporating Grad-CAM into our CNN model, visual evidence of the regions that aided in the classification process is obtained as seen in Figure 3.6. This not only helps in trust building among clinicians and radiologists but also aids in understanding the reasons why the model fails in some cases and how can the model be further improved.

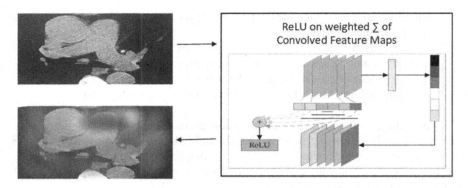

FIGURE 3.6   Grad CAM visualization of an HRCT image with pulmonary arterial hypertension.

Due to limited dataset availability, the sensitivity analysis can't yet be incorporated into the model. In the future, the sensitivity of the model to artery size variation can also be quantified by systematically varying the size of the artery in the input images and observing the corresponding changes in the model's output. While Grad-CAM didn't directly contribute to model accuracy, it plays a crucial role in the interpretability of the model's decision.

3.6   LIMITATIONS AND CHALLENGES OF XAI IN BIOMEDICAL APPLICATIONS

Despite its numerous advantages, XAI in the biomedical world faces certain limitations and challenges.

3.6.1   COMPLEXITY-INTERPRETABILITY TRADEOFF

Achieving high interpretability often comes at the cost of complexity and performance. AI models with increased interpretability may sacrifice predictive accuracy, hindering their practical utility. Balancing interpretability and performance remains a challenge in the development of XAI frameworks for biomedical applications.

3.6.2   COMPLEXITY OF BIOMEDICAL DATA

Biomedical data is often complex, high-dimensional, and multi-modal, posing challenges for explainability. AI models operating on such data may exhibit intricate relationships that are difficult to capture and explain in a human-understandable manner. Developing XAI techniques tailored to the complexity of biomedical data is an ongoing research area.

3.6.3   ETHICAL AND LEGAL CONSIDERATIONS

XAI in biomedical applications raises ethical and legal concerns. The disclosure of sensitive patient information and the potential biases embedded in AI models require careful handling. Striking a balance between transparency and privacy is crucial to address these concerns and ensure the responsible and ethical deployment of XAI in the biomedical domain.

3.7 ANXIETIES AND FUTURE DIRECTIONS

As XAI gains traction in the biomedical world, certain anxieties and challenges need to be addressed.

3.7.1 LACK OF STANDARDIZATION

There is a lack of standardized guidelines and benchmarks for evaluating and comparing different XAI methods in biomedical applications. Establishing common evaluation metrics and datasets can facilitate the development and adoption of XAI techniques, fostering transparency and reproducibility.

3.7.2 REGULATORY HURDLES

Regulatory agencies may struggle to keep pace with rapidly evolving XAI technologies. Establishing regulatory frameworks that ensure the safety, efficacy, and interpretability of AI systems in the biomedical domain is essential. Collaboration between researchers, clinicians, policymakers, and regulatory bodies is crucial to navigate these challenges.

3.7.3 HUMAN-AI INTERACTION

Effective integration of XAI into the clinical workflow requires seamless human-AI interaction. Designing user-friendly interfaces and training programs to help healthcare professionals understand and effectively utilize XAI tools is vital. Ensuring proper education and training can mitigate potential challenges arising from the interaction between humans and AI systems.

3.8 CONCLUSION

Explainable AI (XAI) holds immense potential in the biomedical world, addressing concerns regarding the transparency, trust, and interpretability of AI systems. XAI can revolutionize healthcare delivery and biomedical research through enhanced trust, improved clinical decision-making, and empowering patients. While challenges and anxieties exist, ongoing research, collaboration, and standardization efforts pave the way for a future where XAI plays a central role in shaping biomedical applications.

REFERENCES

Adadi, A., & Berrada, M. (2018). Peeking inside the black-box: A survey on explainable artificial intelligence (XAI). *IEEE Access, 6*, 52138–52160.

Angelov, P. P., Soares, E. A., Jiang, R., Arnold, N. I., & Atkinson, P. M. (2021). Explainable artificial intelligence: An analytical review. *Wiley Interdisciplinary Reviews: Data Mining and Knowledge Discovery*, *11*(5), e1424.

Arrieta, A. B., Díaz-Rodríguez, N., Del Ser, J., Bennetot, A., Tabik, S., Barbado, A., ... & Herrera, F. (2020). Explainable Artificial Intelligence (XAI): Concepts, taxonomies, opportunities and challenges toward responsible AI. *Information Fusion*, *58*, 82–115.

Das, A., & Rad, P. (2020). Opportunities and challenges in explainable artificial intelligence (XAI): A survey. *arXiv Preprint* arXiv:2006.11371.

Deeks, A. (2019). The judicial demand for explainable artificial intelligence. *Columbia Law Review*, *119*(7), 1829–1850.

Došilović, F. K., Brčić, M., & Hlupić, N. (2018, May). Explainable artificial intelligence: A survey. In *2018 41st International Convention on Information and Communication Technology, Electronics and Microelectronics (MIPRO)* (pp. 0210–0215). IEEE.

Jiménez-Luna, J., Grisoni, F., & Schneider, G. (2020). Drug discovery with explainable artificial intelligence. *Nature Machine Intelligence*, *2*(10), 573–584

Langer, M., Oster, D., Speith, T., Hermanns, H., Kästner, L., Schmidt, E., ... & Baum, K. (2021). What do we want from Explainable Artificial Intelligence (XAI)?–A stakeholder perspective on XAI and a conceptual model guiding interdisciplinary XAI research. *Artificial Intelligence*, *296*, 103473.

Minh, D., Wang, H. X., Li, Y. F., & Nguyen, T. N. (2022). Explainable artificial intelligence: A comprehensive review. *Artificial Intelligence Review*, 1–66.

Nazir, S., Dickson, D. M., & Akram, M. U. (2023, Feb). Survey of explainable artificial intelligence techniques for biomedical imaging with deep neural networks. *Computers in Biology and Medicine*, *156*, 106668. DOI: 10.1016/j.compbiomed.2023.106668.

Nazir, A., Naaz, R., Qureshi, S., & Nazir, N. (2022, October). HRCT chest analysis for detection of pulmonary arterial hypertension in COVID-19 patients using convolutional neural networks. In *2022 IEEE 3rd Global Conference for Advancement in Technology (GCAT)* (pp. 1–5). IEEE.

Samek, W., Wiegand, T., & Müller, K. R. (2017). Explainable artificial intelligence: Understanding, visualizing and interpreting deep learning models. *arXiv preprint* arXiv:1708.08296.

Tjoa, E., & Guan, C. (2020). A survey on explainable artificial intelligence (XAI): Toward medical XAI *IEEE transactions on Neural Networks and Learning Systems*, *32*(11), 4793–4813.

Vilone, G., & Longo, L. (2021). Notions of explainability and evaluation approaches for explainable artificial intelligence. *Information Fusion*, *76*, 89–106.

4 XAI in Drug Discovery

Kevser Kübra Kırboğa

4.1 INTRODUCTION

The procedure necessary to create and sell novel drugs is known as drug discovery. Identifying illness etiology, biological targets, possible therapeutic candidates, mechanisms of action, side effects, and clinical results are just a few of the numerous procedures and difficulties involved in drug discovery. To improve the effectiveness and speed of the drug development process, artificial intelligence (AI) techniques are being investigated (BCG, 2022). Target identification, drug design, drug optimization, drug repositioning, and drug toxicity prediction are just a few of the activities in drug research where AI may be applied (Sheu & Pardeshi, 2022).

AI models can be effective in finding new drugs, but they frequently fall short of providing a clear and understandable explanation of how decisions are made. As a result, people may be less inclined to trust and accept AI models (Takagi, Kamada, Hamatani, Kojima, & Okuno, 2022). In terms of medicinal relevance, ethical norms, and laws, the explainability of AI models in drug development is equally crucial (Jiménez-Luna, Grisoni, & Schneider, 2020). To better understand how AI models function in drug development and why they yield specific findings, researchers are paying increasing attention to explainable artificial intelligence (XAI) (Holzinger, 2020).

The goal of XAI, a subfield of AI, is to use a range of strategies and methodologies to make the decision-making processes of AI models transparent and understandable. By scrutinizing the inputs, outputs, or internal structure of AI models, XAI approaches can produce explanations (Harren, Matter, Hessler, Rarey, & Grebner, 2022; Wang, Huang, Chandak, Zitnik, & Gehlenborg, 2022). On the other hand, XAI methods can offer different approaches to increase the explainability of AI models, for example, knowledge-based methods, model distillation methods, or interpretable machine learning methods (Ghassemi et al., 2020). XAI techniques and methods can increase AI models' biomedical relevance, reliability, and accountability in drug discovery (Carey & Papin, 2018).

In this chapter, we aim to review XAI's applications, techniques, and challenges in drug discovery. We also discuss future directions and opportunities for XAI in this area. The contribution of this section is as follows:

DOI: 10.1201/9781003426073-4

- We highlight the importance and benefits of XAI in drug discovery.
- We present the application areas and examples of XAI in drug discovery.
- We describe XAI techniques and methods used in drug discovery.
- We outline XAI challenges and solutions in drug discovery.
- We propose future directions and opportunities for XAI in drug discovery.

Here's how we proceed in the remainder of the chapter: Section 2 explains the importance and benefits of XAI in drug discovery. Section 3 presents the application areas and examples of XAI in drug discovery. Section 4 describes the XAI techniques and methods used in drug discovery. Section 5 outlines the XAI challenges and solutions in drug discovery. Section 6 proposes future directions and opportunities for XAI in drug discovery. We summarize the results in section 7.

4.2   THE IMPORTANCE AND BENEFITS OF XAI IN DRUG DISCOVERY

XAI is a field of research that can explain the decisions and workings of AI systems in an understandable way to humans. XAI aims to increase the reliability, accountability, and ethics of AI applications (K. K. Kırboğa, Abbasi, & Küçüksille). Drug discovery and development is a long, complex, costly process of finding, testing, and bringing new drug candidates to market (Figure 4.1). In this process, AI techniques are used in various stages, such as data analysis, molecular modelling, biological simulation, and clinical trial design (Askr et al., 2023). However, the ability of AI systems to explain how and why their decisions are made is an essential requirement in drug discovery and development because these systems produce false or misleading results, which can harm human health (Manresa-Yee, Roig-Maimó, Ramis, & Mas-Sansó, 2022). XAI ensures the transparency and reliability of AI systems in drug discovery and development. Thanks to XAI, AI systems:

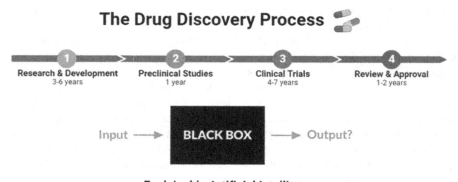

The Drug Discovery Process

1 Research & Development
3-6 years

2 Preclinical Studies
1 year

3 Clinical Trials
4-7 years

4 Review & Approval
1-2 years

Input ⟶ **BLACK BOX** ⟶ Output?

Explainable Artificial Intelligence

FIGURE 4.1   With Explainable Artificial Intelligence, the drug discovery process, which is costly, complex and consists of specific phases, can be transformed into an explainable and effective method with less cost and time.

- can control the quality and integrity of the data;
- explain the properties and interactions of molecular structures;
- interpret the functioning of biological mechanisms;
- evaluate the results of clinical trials; and
- predict the efficacy and safety of drug candidates (Manresa-Yee et al., 2022).

Many studies in different languages on the importance of using XAI in drug discovery confirm these data. Drance et al. present the neurosymbolic XAI applied to drug repositioning for rare diseases. It performs drug repositioning using link prediction algorithms and explains how and why these algorithms produce results. It also proposes a method for understanding how data organisation in the knowledge graph changes the forecast quality. This study demonstrates that XAI enables transparency and innovation in drug repositioning (Drancé, 2022). Askr et al. presents a systematic literature review on deep learning-based drug discovery and development approaches. It reviews more than 300 articles between 2000 and 2022 and presents critical datasets, databases, and evaluation criteria. In addition, this study provides an overview of how XAI supports drug discovery issues and discusses drug dosage optimization and success stories. Finally, digital twinning (DT) and open issues are suggested as future research challenges for drug discovery problems. This study demonstrates that XAI improves performance and intelligibility in drug discovery (Askr et al., 2023). Manresa-Yee et al. provide an overview of the contexts in which annotation interfaces are used in healthcare. It compiles health-related studies by searching major research databases and shows that researchers offer explanations in the form of natural text, parameter effects, data graphs, or salience maps. This study shows that XAI improves user confidence and healthcare decision quality (Manresa-Yee et al., 2022).

4.3 TECHNIQUES AND METHODS USED IN DRUG DISCOVERY OF XAI

The applications of XAI in drug discovery span many disciplines. Target identification is one of drug discovery's most important and challenging steps. A target is a molecule or biological structure that affects the course or symptoms of a disease. AI systems can analyze large and complex data sources in target identification. For example, omics data (genomics, proteomics, metabolomics, etc.) can be used to reveal the molecular mechanisms of diseases. Text data (scientific articles, patents, clinical reports, etc.) can provide information about the functions and relationships of targets. Image data (medical images, microscope images, crystal structures, etc.) can provide information about the morphology and localization of targets. AI systems can identify new and potential targets by integrating these data sources. Rasul et al. conducted a study on choosing machine learning methods for target identification (Rasul et al., 2022). This study showed that databases of drug-target interactions were statistically biased, thus leading to false positive estimates. To solve this problem, he proposed a new negative sample selection scheme that ensures that each protein and each drug is included in equal numbers in positive and negative samples. It has been shown that machine learning models trained with this scheme produce fewer false positive predictions in target identification and improve the sequencing of correct targets. Najm et al. provided an overview of the various methods used in target identification (Najm, Azencott, Playe, & Stoven, 2021). In this study, it was stated that the methods used in target identification were classified as affinity methods, genetic methods, computational methods, and chemical proteomics. In addition, these methods' working principles, application areas, and advantages and disadvantages are explained. De novo molecular design is one of the most creative and innovative stages in drug discovery. De novo molecular design is a method used to design new chemical compounds independently of existing molecules. AI systems can create novel chemical substances by examining the characteristics and interactions of molecular structures in de novo molecular design. Deep learning systems, for instance, may create novel compounds by either visually portraying or recording molecular structures in chemical languages (such as SMILES). Substances' physicochemical and pharmacokinetic features can be optimized using these algorithms to provide the desired biological activity. AI systems can use these methods to create fresh and potent medication candidates. A novel AI technique for de novo molecular design was created by Liu et al. (2020). This paper proposes a deep learning approach that employs graphical neural networks to describe chemical structures visually. This approach can create novel molecules by optimising both chemical characteristics and biological activity. This model also accounts for variety and synthesability. Prykhodko et al. allegedly created a new AI technique for de novo molecular design (Prykhodko et al., 2019). This work suggests a deep learning model based on transformative learning to describe chemical structures using SMILES coding. This model can produce novel molecules by using methods like data augmentation, pre-training, and fine-tuning. This approach can also optimise the physicochemical and pharmacokinetic features of the substances to provide the desired biological activity. Drug dose optimisation is one of the most crucial and delicate phases of medication research. Medication dosage optimization is a technique for modifying a patient's medication dosage to maximize their response to therapy. In medicine dosage optimization, AI systems may select the most suitable drug dose based on the patient's specific information (e.g., age, gender, weight, genetics, metabolism, illness condition, etc.). For instance, reinforcement learning algorithms can dynamically change medicine doses based on how patients respond to therapy. These algorithms can adjust a drug's dose to improve its efficacy and lessen its negative effects. AI systems can use these algorithms to enhance patients' health and quality of life. According to Sehgal et al., a novel AI technique for optimizing medicine doses has been created (Sehgal, Ward, La, & Louis, 2022). A deep learning model based on genetic algorithms is suggested in this paper. This model can optimize the medicine dosage using the patient's genetic information. This model may adjust the drug dosage by accounting for pharmacodynamics and metabolism. In a different research, a novel approach is put out that stresses the significance of explaining deep learning architectures' decision-making processes in drug development. The authors automatically recognized and extracted different deep representations from 1D sequential and structural data using convolutional neural networks (CNN). CNNs have proven effective at locating binding sites and other key binding sites. They also identified the input areas that contributed most to the model's estimation, which helped to explain the model's choices. This technique fared better than the alternatives in estimating and sequencing binding interaction strength. This work supports the possible use of an end-to-end deep learning architecture in drug development outside the restricted domain of 3D structured proteins and ligands. Explaining the decision-making process also proves the validity of deep representations extrapolated from CNNs (Monteiro et al., 2022). To predict bond interaction intensity and clarify the meaning of the deep representations CNNs infer, this work suggests an end-to-end deep learning architecture that uses 1D sequential and structured data. In the area of drug development, this technique enhances both data sources and model intelligibility.

4.4   TECHNIQUES AND METHODS USED IN DRUG DISCOVERY OF XAI

The techniques and tools employed by XAI in drug discovery include annotation models, interfaces, and visualization tools. The findings or characteristics generated by AI systems can be graphically expressed using visualization techniques. These methods can display molecular interactions or attributes using various visual representations of molecules, such as 2D or 3D diagrams, SMILES arrays, or infographics. Clarification interfaces are tools for displaying the justifications or logic behind the choices made by AI systems (Heberle, Zhao, Schmidt, Wolf, & Heinrich, 2023). Users can enter any score assigned to atomic and non-atomic tokens and visualize them on top of a 2D molecular diagram in coordination with a bar graph representing a string of SMILES (K. Kırboğa, Kucuksille, & Köse, 2022). Dan et al. introduce a new molecule generation method that uses a graph-based generative adversarial network (GAN) to generate molecules with desired properties (Dan et al., 2019). This method uses graph convolution to encode molecules and graph deconvolution to decode hidden vectors into molecule graphs. Gómez-Bombarelli et al. present a deep learning-based method for designing molecules with desired pharmacological properties (Gómez-Bombarelli et al., 2016). This method uses a recurrent neural network to generate SMILES sequences representing molecules and a convolutional neural network to evaluate the properties of the produced molecules. Jin et al. present a new deep learning-based molecular design method to produce new molecular structures according to the target structure and activity conditions (Jin, Yang, Barzilay, & Jaakkola, 2018). This method uses a variational autoencoder and reinforcement learning-based framework that can optimize the similarity of the molecular structure and the activity score while ensuring the plausibility of the molecular structure.

4.5   CHALLENGES AND SOLUTIONS IN DRUG DISCOVERY OF XAI

XAI can be defined as a field that aims to make the decisions and operations of AI systems understandable for humans. XAI is essential for increasing AI applications' reliability, transparency, and accountability in complex and critical areas such as drug discovery (Askr et al., 2023) (Figure 4.2). Drug discovery and development are essential translational science activities contributing to human health and well-being. However, developing a new drug is a rather complex, expensive and lengthy process

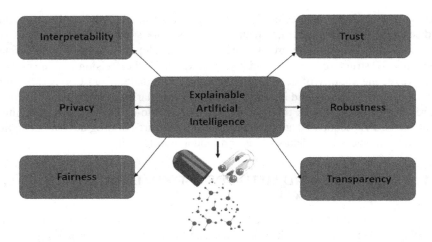

FIGURE 4.2 The characteristics of Explainable Artificial Intelligence that affect the drug discovery process.

(Antoniadi et al., 2021). How to reduce costs and accelerate new drug discovery has become a challenging and urgent question in the industry (Antoniadi et al., 2021). Combining AI with new experimental technologies is expected to make the search for new drugs faster, cheaper, and more effective (Antoniadi et al., 2021). With the widespread use of AI in drug discovery and development, XAI seems to play an important role in overcoming the challenges in this field. XAI offers several methods to explain how AI models used in the drug discovery process work, what data it uses, what features it selects, what results it produces, and how reliable these results are (Askr et al., 2023). In this way, it is possible to evaluate the performance of AI models, detect errors, improve and optimize them. In addition, thanks to XAI, it is possible to consider AI models' ethical, legal, and social responsibilities (Askr et al., 2023). In the initial phase of drug discovery, potential molecular targets are identified that can ameliorate the biological disorder caused by the disease (Antoniadi et al., 2021). At this stage, XAI can be used to check the quality, integrity, and compatibility of data used by AI models in target selection (Askr et al., 2023). In addition, XAI can be used to describe which features AI models use in target selection and the degree of importance of these features (Askr et al., 2023). For example, a DL model called DeepDTA makes DTI (Drug-Target Interactions) predictions using protein-sequence data (Manresa-Yee et al., 2022). The XAI version of this model, DeepDTA-X, uses a model-specific method called Grad-CAM to explain its predictions. Using this method, it is shown which protein regions the model focuses on and how these regions contribute to the prediction. In the second step of drug discovery, small molecules (ligands) that can bind to the selected target are designed or found by screening methods (Askr et al., 2023). The quality, integrity, and compatibility of the data utilized by AI models in ligand design can now be verified using XAI (Askr et al., 2023). The properties that AI models utilize in ligand design and their significance may also be described using XAI (Askr et al., 2023). Additionally, XAI can show the confidence interval, margin of error, and success percentage of the ligand design findings generated by AI models (Askr et al., 2023). For example, a DL model called DeepPurpose designs ligands using molecular structure data. The XAI version of this model, DeepPurpose-X, uses a model-independent method called SHAP to explain its predictions. This method shows which molecular features the model focuses on and how these features contribute to the prediction. In the third phase of drug discovery, the pharmacokinetic, pharmacodynamic, toxicological, and metabolic properties of the selected ligands are optimized (Antoniadi et al., 2021). XAI can be used to check the quality, integrity, and compatibility of data used by AI models in ligand optimization. In addition, XAI can be used to describe which features AI models use in ligand optimization and the importance of these features. XAI can also display the confidence interval, margin of error, and success rate of the results produced by the AI models in ligand optimization (Askr et al., 2023). For example, a DL model called Mol-CycleGAN performs ligand optimization using molecular structure data. Mol-CycleGAN-X, the XAI version of this model, uses a model-independent method called LIME to explain its predictions. Using this method, it is shown which molecular properties the model changes and how these changes contribute to the prediction (Maziarka et al., 2020).

4.6 XAI FUTURE DIRECTIONS AND OPPORTUNITIES IN DRUG DISCOVERY

XAI is an approach that presents the reasons and logic of the outputs produced by AI systems to the users. XAI is especially important in applications where users need to interpret the outputs of AI systems. XAI is also an area with great potential in drug discovery. In drug discovery, machine learning and graphic neural networks show promising results in molecular design, chemical synthesis planning, and molecular property prediction tasks. For specialists in drug development, many of these techniques are viewed as a "black box" since they are difficult to understand. As a result, XAI is crucial for making moral, impartial, and trustworthy decisions in drug development. Systems that help clinicians make clinical decisions include XAI applications for Clinical Decision Support Systems (CDSS). Making these systems understandable will help professionals make more thoughtful and occasionally life-saving judgments. A methodical literature study investigates present issues and potential future developments in the CDSS domain of XAI (Antoniadi et al., 2021). When applying XAI for CDSS, several XAI strategies should be selected in line with user demands and circumstances. Local descriptions, for instance, might aid users in decision-making in a specific circumstance, but global descriptions can aid users in gaining perspective or assessing the system's dependability. Application of XAI for CDSS also requires consideration of evaluation techniques. Evaluation techniques should include both quantitative and qualitative criteria to gauge the efficiency, dependability, and usefulness of XAI. The requirements and expectations of various user groups (such as physicians, nurses, and patients) should also be considered while developing evaluation methodologies. The use of XAI for CDSS should consider ethical and legal obligations. By encouraging user engagement in the decision-making process, XAI can be helpful for moral and just decision-making. However, XAI must also safeguard users' consent, security, and privacy.

Additionally, XAI has to stress that users are accountable for the results of their choices. The Explainable Graphical Neural Networks (GNN) for the Molecular Property Prediction approach uses explainable graphic neural networks to train on learnt representations of molecules (Rao, Zheng, Lu, & Yang, 2022). This method shows high performance in molecular property prediction, an essential task in drug discovery. However, the explainability of GNNs is limited and based on people's judgments. In this study, a three-level comparison dataset was created to evaluate the explainability of GNNs quantitatively. In addition, recently developed XAI methods and different GNN algorithms are applied, and their benefits, limitations, and future opportunities for drug discovery are highlighted (Rao et al., 2022). To assess the explainability of GNNs, quantitative and qualitative criteria should be used. Quantitative criteria can be used to measure disclosures' accuracy, consistency, and comparability. Qualitative criteria can be used to measure the understandability, usefulness, and user satisfaction of explanations. To assess the explainability of GNNs, different datasets should be used. Different datasets should contain different graph sizes, degree distributions, and homophilic or heterophilic graphs. Also, datasets must be resistant to the known pitfalls of explainable algorithms. For example, labels in the dataset should not have more than one root cause, or there should be cases where a random node or edge is not sufficient as an explanation. To assess the explainability of GNNs, different XAI methods should be used. Different XAI methods may provide different annotation types (e.g., local or global, model-specific or model-agnostic) and have different advantages and disadvantages. For example, gradient-based methods such as GradInput and IG may provide the best model interpretability for GNNs. In contrast, perturbation-based methods such as LIME and SHAP may have higher computational costs (Jiménez-Luna, Grisoni, Weskamp, & Schneider, 2021).

4.7 CONCLUSION

XAI makes the decisions and operations of AI models used in drug discovery and development understandable to humans, thus increasing the reliability, transparency, and accountability of AI applications in this field. In this way, it is possible to evaluate the performance of AI models, detect errors, and improve and optimize them. In addition, thanks to XAI, it is possible to consider AI models' ethical, legal, and social responsibilities. XAI develops various methods that offer solutions to drug discovery and development challenges. These methods describe what data AI models use in target selection, ligand design, and ligand optimization, what features they select, what results they produce, and how reliable they are. These methods can be model-specific or model-independent, local or global, antecedent. XAI uses different interfaces to make the decisions and operations of AI models understandable in drug discovery and development. These interfaces may include visual or textual elements to show which features the AI models focus on and the contribution of those features to the prediction. These interfaces can also allow users to give feedback to AI models or adjust parameters of AI models.

REFERENCES

Antoniadi, A. M., Du, Y., Guendouz, Y., Wei, L., Mazo, C., Becker, B. A., & Mooney, C. (2021). Current challenges and future opportunities for XAI in machine learning-based clinical decision support systems: A systematic review. *Applied Sciences, 11*(11), 5088. Retrieved from https://www.mdpi.com/2076-3417/11/11/5088

Askr, H., Elgeldawi, E., Aboul Ella, H., Elshaier, Y. A. M. M., Gomaa, M. M., & Hassanien, A. E. (2023). Deep learning in drug discovery: An integrative review and future challenges. *Artificial Intelligence Review, 56*(7), 5975–6037. doi:10.1007/s10462-022-10306-1

BCG. (2022). Adopting AI in Drug Discovery. Retrieved from www.bcg.com/publications/2022/adopting-ai-in-pharmaceutical-discovery

Carey, M. A., & Papin, J. A. (2018). Ten simple rules for biologists learning to program. *PLOS Computational Biology, 14*(1), e1005871. doi:10.1371/journal.pcbi.1005871

Dan, Y., Zhao, Y., Li, X., Li, S., Hu, M., & Hu, J. (2020). Generative adversarial networks (GAN) based efficient sampling of chemical composition space for inverse design of inorganic materials. *npj Computational Materials, 6*(1), 84. https://doi.org/10.1038/s41524-020-00352-0

Drancé, M. (2022). *Neuro-Symbolic XAI: Application to Drug Repurposing for Rare Diseases.* Paper Presented at the Database Systems for Advanced Applications: 27th International Conference, DASFAA 2022, Virtual Event, April 11–14, 2022, Proceedings, Part III. https://doi.org/10.1007/978-3-031-00129-1_51

Ghassemi, M., Naumann, T., Schulam, P., Beam, A. L., Chen, I. Y., & Ranganath, R. (2020). A review of challenges and opportunities in machine learning for health. *AMIA Joint Summits on Translational Science Proceedings, 2020*, 191–200.

Gómez-Bombarelli, R., Duvenaud, D., Hernández-Lobato, J., Aguilera-Iparraguirre, J., Hirzel, T., Adams, R., & Aspuru-Guzik, A. (2016). Automatic chemical design using a data-driven continuous representation of molecules. *ACS Central Science, 4*. doi:10.1021/acscentsci.7b00572

Harren, T., Matter, H., Hessler, G., Rarey, M., & Grebner, C. (2022). Interpretation of structure-activity relationships in real-world drug design data sets using explainable artificial intelligence. Journal of Chemical Information and Modeling, *62*(3), 447–462. doi:10.1021/acs.jcim.1c01263

Heberle, H., Zhao, L., Schmidt, S., Wolf, T., & Heinrich, J. (2023). XSMILES: Interactive visualization for molecules, SMILES and XAI attribution scores. *Journal of Cheminformatics, 15*. doi:10.1186/s13321-022-00673-w

Holzinger, A. (2020). Explainable AI and multi-modal causability in medicine. *i-com, 19*(3), 171–179. doi:doi:10.1515/icom-2020-0024

Jiménez-Luna, J., Grisoni, F., & Schneider, G. (2020). Drug discovery with explainable artificial intelligence. *Nature Machine Intelligence, 2*(10), 573–584. doi:10.1038/s42256-020-00236-4

Jiménez-Luna, J., Grisoni, F., Weskamp, N., & Schneider, G. (2021). Artificial intelligence in drug discovery: recent advances and future perspectives. *Expert Opinion on Drug Discovery, 16*(9), 949–959. doi:10.1080/17460441.2021.1909567

Jin, W, Yang, K, Barzilay, R and Jaakkola, T. (2019). Learning multimodal grapht o-graph translation for molecular optimization. 7th International Conference on Learning Representations, ICLR 2019.

Kırboğa, K. K., Küçüksille, E. U., & Utku K. (2023). Ignition of Small Molecule Inhibitors in Friedreich's Ataxia with Explainable Artificial Intelligence. BRAIN. Broad Research in Artificial Intelligence and Neuroscience. Print.

Kırboğa, K. K., Abbasi, S., & Küçüksille, E. U. (2023). Explainability and white box in drug discovery. *Chemical Biology & Drug Design*, 102(1), 217–233. https://doi.org/10.1111/cbdd.14262

Liu, C., Korablyov, M., Jastrzębski, S., Bengio, Y., & Segler, M. H. (2020). RetroGNN: Approximating Retrosynthesis by Graph Neural Networks for de novo drug design. *ArXiv*. /abs/2011.13042

Manresa-Yee, C., Roig-Maimó, M. F., Ramis, S., & Mas-Sansó, R. (2022). Advances in XAI: explanation interfaces in healthcare. In C.-P. Lim, Y.-W. Chen, A. Vaidya, C. Mahorkar, & L. C. Jain (Eds.), *Handbook of Artificial Intelligence in Healthcare: Vol 2: Practicalities and Prospects* (pp. 357–369). Cham: Springer International Publishing.

Maziarka, Ł., Pocha, A., Kaczmarczyk, J., Rataj, K., Danel, T., & Warchoł, M. (2020). Mol-CycleGAN: A generative model for molecular optimization. *Journal of Cheminformatics, 12*(1), 2. doi:10.1186/s13321-019-0404-1

Monteiro, N. R. C., Simões, C. J. V., Ávila, H. V., Abbasi, M., Oliveira, J. L., & Arrais, J. P. (2022). Explainable deep drug-target representations for binding affinity prediction. *BMC Bioinformatics, 23*(1), 237. doi:10.1186/s12859-022-04767-y

Najm, M., Azencott, C.-A., Playe, B., & Stoven, V. (2021). Drug target identification with machine learning: how to choose negative examples. *International Journal of Molecular Sciences, 22*(10), 5118. Retrieved from www.mdpi.com/1422-0067/22/10/5118

Prykhodko, O., Johansson, S. V., Kotsias, P.-C., Arús-Pous, J., Bjerrum, E. J., Engkvist, O., & Chen, H. (2019). A de novo molecular generation method using latent vector based generative adversarial network. *Journal of Cheminformatics, 11*(1), 74. doi:10.1186/s13321-019-0397-9

Rao, J., Zheng, S., Lu, Y., & Yang, Y. (2022). Quantitative evaluation of explainable graph neural networks for molecular property prediction. *Patterns (NY), 3*(12), 100628. doi:10.1016/j.patter.2022.100628

Rasul, A., Riaz, A., Sarfraz, I., Khan, S. G., Hussain, G., Zara, R., Sadiqa, A., Bushra, G., Riaz, S., Iqbal, M. J., Hassan, M., & Khorsandi, K. (2022). Target Identification Approaches in Drug Discovery. In M. T. Scotti & C. L. Bellera (Eds.), Drug Target Selection and Validation (pp. 41–59). Springer International Publishing. https://doi.org/10.1007/978-3-030-95895-4_3

Sehgal, A., Ward, N., La, H. M., & Louis, S. J. (2022). Automatic parameter optimization using genetic algorithm in deep reinforcement learning for robotic manipulation tasks. *ArXiv*, abs/2204.03656.

Sheu, R.-K., & Pardeshi, M. S. (2022). A survey on medical explainable AI (XAI): Recent progress, explainability approach, human interaction and scoring system. *Sensors, 22*(20), 8068. Retrieved from www.mdpi.com/1424-8220/22/20/8068

Takagi, A., Kamada, M., Hamatani, E., Kojima, R., & Okuno, Y. (2022). GraphIX: Graph-based In silico XAI (explainable artificial intelligence) for drug repositioning from biopharmaceutical network. *ArXiv*, abs/2212.10788.

Wang, Q., Huang, K., Chandak, P., Zitnik, M., & Gehlenborg, N. (2023). Extending the Nested Model for User-Centric XAI: A design study on GNN-based drug repurposing. *IEEE Transactions on Visualization and Computer Graphics, 29*(1), 1266–1276. https://doi.org/10.1109/tvcg.2022.3209435

5 The Use of Explainable Artificial Intelligence in Medical Image Processing
A Research Study

*Remzi Gürfidan, Bekir Aksoy, Mevlüt Ersoy,
Enes Açıkgözoğlu, and Sema Çayır*

5.1 MEDICAL IMAGING METHODS

As the development of technology affects all sectors, it also makes great contributions to the field of health. One of the most important contributions is undoubtedly the imaging methods developed for the diagnosis and treatment of patients. We briefly explain these methods. X-ray imaging is a technique that uses X-rays to show the body's internal anatomy. Numerous illnesses, including fractures, lung infections, and the detection of cancer, are diagnosed with it. With the help of powerful magnetic fields and radio waves, magnetic resonance imaging (MRI) creates cross-sectional images of the human body (Fessler, 2010). Visualizing soft tissues, organs, the nervous system, and blood vessels in detail is beneficial. Using X-rays, a technique known as computed tomography (CT) creates incredibly tiny cross-sectional pictures. The cross-sectional images can be combined to provide three-dimensional images (Withers et al., 2021). It is used to evaluate many bodily parts, including the head, chest, and belly. High-frequency sound waves are used in ultrasonography to visualize the inside anatomy of the body. It is utilized in a variety of contexts, including tumour localization, organ evaluation, and pregnancy monitoring. Positron Emission Tomography (PET) is a technique that (Bailey et al., 2005) uses positrons released by tracer compounds injected into the body by a radioactive injection to assess metabolism and functional activity. It is used to diagnose and monitor neurological problems, heart disease, and cancer. Mammography is a screening and diagnosis tool for breast cancer based on X-rays (Gøtzsche & Jørgensen, 2013). It aids in observing breast tissue and locating any anomalies.

In the field of medical imaging, these techniques are the most often employed. Each technique offers a unique set of data that is useful for diagnosing and tracking

DOI: 10.1201/9781003426073-5

diseases. Medical imaging techniques are constantly being developed and enhanced because to developments in science and technology.

5.2 EXPLAINABLE ARTIFICIAL INTELLIGENCE TECHNOLOGY

Explainable AI is a technology that enables humans to comprehend and trace the decisions or results of an AI system. Explainable AI employs more transparent and intelligible approaches to explain decision processes and consequences, in contrast to typical AI models, which can frequently be complex and difficult to understand (van der Velden et al., 2022; Tjoa & Guan, 2021). Explainable AI is frequently distinguished by its capacity to comprehend decision-making processes and results, identify faults, clear up misunderstandings, or verify that systems adhere to ethical and legal standards. People can trust the decisions made by AI systems when they are able to explain their reasoning.

Explainable artificial intelligence can be achieved using a variety of techniques and strategies. By incorporating feature weights or the significance of features, some techniques offer simple explanations. Other methods favor easily understandable forms such as cause-and-effect chains, logic-based rules or decision trees. Explaining elements like the dataset and training procedure used by AI systems is also crucial. When utilized in vital systems, such as healthcare, financial decision-making, or judicial systems, explainable artificial intelligence (XAI) is especially intriguing. It is simpler for people to trust these systems and defend the rights and safety of users in these areas because decisions and results are explicable. The popular and widely used XAI algorithms are briefly described here.

- The explainability technique LIME (Local Interpretable Model-Agnostic Explanations) is used to explain the predictions of machine learning models. The model-agnostic approach of LIME makes it a well-known method for discussing complex models. It is compatible with all machine learning models because of this property (Nagaraj et al., 2022).
- SHAP (Shapley Additive Explanations), an explainability technique, is used to compute feature weights and provide explanations for predictions. SHAP works by estimating each feature's contribution using the cooperative game theory concepts, then calculating Shapley values to describe this contribution (Roshan & Zafar, 2022).
- L2X (Layer-wise Relevance Propagation) is a technique for enhancing the explainability of models based on neural networks. The contribution of each input feature to the prediction outcome is ascertained through backpropagation in this method. L2X demonstrates which input elements have a greater influence on the outcome of the prediction (Jung et al., 2021).
- DeepLIFT (Deep Learning Important FeaTures) is an explainability technique used to pinpoint key characteristics in models based on neural networks. Backpropagating gradients are used in this technique to determine how much an input feature contributed to the outcome of the prediction. DeepLIFT explains how each factor affects the outcome of the forecast.

- Anchor is a method for describable AI models on text data. This method explains text predictions by creating small rules that are conceptually simple and understandable. Anchor annotates by identifying important features and conceptual connections in texts (Jouis et al., 2021).
- Grad-CAM (Gradient-weighted Class Activation Mapping) is a method used to explain the visual results of neural network-based models. This method uses gradient information to explain the classification result and show which regions of the image contribute more to the prediction result (Banerjee et al., 2022).

Each algorithm offers a different approach and explainability method. The algorithm to be chosen depends on the type of data to be used, the model structure, and the targeted explainability criteria.

5.3 PROCESSING MEDICAL IMAGES WITH EXPLAINABLE ARTIFICIAL INTELLIGENCE

X-ray Images, Magnetic Resonance Imaging (MRI), Computed Tomography (CT), and Ultrasound Images are among the most preferred medical imaging modalities in the healthcare industry. Explainable artificial intelligence methods can be used with various types of medical images. By using different analysis and annotation methods for each image type, XAI models can be developed for many purposes such as pathology detection, classification, and region identification. Several technical steps should be performed to process medical images with XAI, analyze the images, and explain the results in an understandable way. Although these technical steps vary depending on the nature and purpose of the study, they should be carried out in a certain order and process within the framework of general rules.

First of all, a dataset consisting of the medical images to be used should be prepared. The prepared dataset must be correctly labelled. Deep learning methods (Convolutional Neural Networks, CNNs) are usually used for training. This model is capable of learning to recognize and classify different pathologies or anatomical structures in medical images. Explainability methods must be applied to explain the results of the trained model. These techniques try to provide a clear explanation of the model's decision-making and output. Explainability techniques demonstrate how the model responds to a specific image and how the output is affected. These details are used to provide explanations for the outcomes of the medical image analysis and the retrieved features. A validation method is carried out to score the generated explanations' accuracy and efficacy. The annotations accurately portray the situation as it is, according to experts.

Doctors and other healthcare workers would benefit greatly from the processing of medical images using XAI in the diagnosis, diagnosis, and treatment processes. By making decision-making for patients and hospital administration transparent, it can help boost trust in authority. It's crucial to keep in mind that XAI model applications should properly preserve security, privacy, and ethical concerns.

5.4   THE IMPORTANCE OF EXPLAINABLE ARTIFICIAL INTELLIGENCE IN MEDICAL IMAGE PROCESSING

It is necessary to examine the role of XAI in medical image processing from several different perspectives. Analysing medical images accurately and ensuring that the results are reliable is critical for making the right diagnostic and treatment decisions. Explainable AI methods can explain the model's decision processes and outcomes so that healthcare professionals and patients understand why the model gives a certain result and has a high level of confidence in the decisions. In addition, it explains the internal working mechanisms of complex AI models in an understandable way. It shows how the model makes decisions and which features influence the results. Healthcare professionals can monitor the decision-making process and outcomes, diagnose errors, and understand the factors that influence outcomes. This is an indirect contribution. This contribution is to provide education based on cause and effect relationships to the students of the department in the training process of health professionals and specialists. If we look at the whole process from the patients' perspective, patients will be able to easily understand their own health status and images. With explainability, patients can better understand pathologies or structures in their images and make more informed health decisions. Patients can participate more actively in treatment processes by trusting the results presented by explainable AI. For all these reasons, XAI in medical image processing plays an important role in healthcare. Explainability ensures that accurate diagnoses are made, treatment decisions are more grounded, reliability and trust are established, and patients are better informed.

In the processing of medical images, XAI aims to provide reliable diagnosis and treatment with its mission, to enable healthcare professionals and patients to make informed decisions with transparency and understandability, and to achieve human-AI collaboration, continuous learning and development, ethical compliance, and compliance with legal norms with its vision.

5.5   ACADEMIC STUDIES IN MEDICAL IMAGE PROCESSING

A total of 232 articles, 70 proceeding papers, 35 review articles, 3 book chapters, and 1 abstract were published between 2018 and 2023 on the processing of medical images with XAI. Sixty-four of these publications are on Springer Nature, 58 Elsevier, 55 IEEE, 53 MDPI, and 15 Wiley platforms. According to the results obtained from the search in the Web of Science search engine, the keyword links of the studies containing the keywords XAI and Medical Image and their colored form according to the years are shown in Figure 5.1.

According to the results obtained from the Web of Science search engine, the heat map of the studies containing the keywords XAI and Medical Image is shown in Figure 5.2.

Jahmunah et al. studied the application of XAI for accurate electrocardiography (ECG) prediction in emergency myocardial infarction (MI), which is responsible for a high number of deaths worldwide. DenseNet and CNN models were developed to classify healthy subjects and patients with ten classes of MI based on the location of

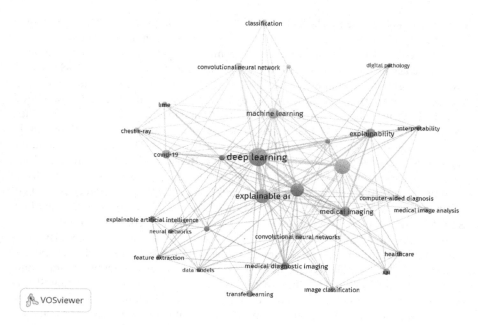

FIGURE 5.1   Keywords associated with XAI in medical image processing studies by year.

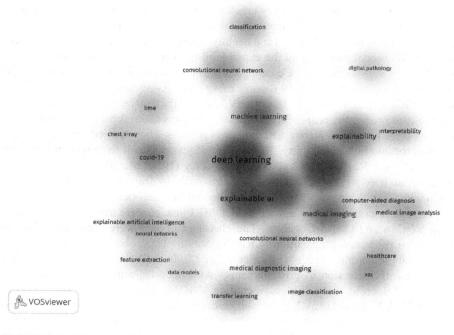

FIGURE 5.2   Heatmap of keywords associated with XAI in medical image processing studies by year.

myocardial involvement. The Grad-CAM activation mapping (CAM) methodology was the best technique for the models' predictions. This technique was used to visualize particular ECG leads and ECG wave segments from the outputs of both models. Average classification accuracy for the DenseNet and CNN models was 98.9% and 98.5%, respectively (Jahmunah et al., 2022). Al Hammadi et al. presented an insider risk assessment system as a fitness-for-duty safety assessment using EEG brainwave signals. They identified incoming threats using AI and machine learning methods that can explain the security of EEG signals. A dataset was created from 17 different individuals using a five-electrode Emotiv Insight EEG device. The data were classified with two- and one-dimensional CNN, Adaptive Boosting, Random forest, and K-nearest neighbor models. Classification results were 96%, 75%, 97%, 94%, and 81% accuracy, respectively (Al Hammadi et al., 2021). In their study, Anand et al. applied automatic analysis of public ECG signals with AI in CNN algorithm. The applied ST-CNN-GAP-5 model was observed to give better results compared to other technologies. An Area Under the Curve (AUC) value of 93.4% was obtained. By generalizing the deep learning model for ECG datasets, the same network architecture was evaluated on another ECG dataset of arrhythmia patients, yielding accuracy and AUC results of 95.8% and 99.46%, respectively. We analyzed ECG data using SHAP on trained ST-CNN-GAP-5 in order to assess the explainability or interpretability of the conclusions reached by this deep convolutional network model. The outcome demonstrates that the model can highlight pertinent ECG wave alterations as required by physicians and explaining them for diagnostic reasons (Anand et al., 2022). In their study, Örnek et al. trained and interpreted tomochrophic images of newborn babies with evolved neural networks. They created 190 tomography images taken from 38 different babies hospitalized in the neonatal unit as a dataset. They trained the CNN model to classify them as diseased and healthy. Then, they visualized the output layers of the CNN model and ensured the explainability of the model. Model success was measured as 97.3% (Ornek & Ceylan, 2020). In their study, Hu et al. tried to make a definitive diagnosis of Covid-19 disease using chest X-ray images with CNN. They used VGG16, ResNet152V2, InceptionV3, and EfficientNetB3 models as neural networks to predict the probability of disease. Datasets taken from the internet environment were trained with four different CNN models. They applied the fuzzy filters method for image explainability. The best result was obtained from the combination of VGG16 and ResNet152V2 with 95% accuracy (Hu et al., 2022). In their study, Sudar et al. tried to detect Alzeimer's disease early by training MRI images of the brain with CNN. More than 6500 images collected from an open source network were used as a dataset to train and test the model. MRI images were trained and classified with VGG16 and CNN algorithms. The classified images were annotated with the layer-wise relevance propagation (LRP) algorithm. They obtained an accuracy of 78.1% with the applied model (Sudar et al., 2022). In their study, Hassan et al. tried to detect early detection of prostate cancer with deep learning models. They used pre-trained deep learning models, including VGG16, MobileNetV2, ResNet50V2, ResNet101V2, Resnet152V2, Xception, InceptionResNetV2, and InceptionV3, to retrain ultrasound and MRI images obtained from public sources. By integrating the effectiveness of the pre-trained model with other machine learning algorithms SVM, Adaboost, K-NN, and Random Forests as feature extractors, they created a fusion model. The fusion model was then examined with explainable AI to find out why prostate cancer is detected as benign or malignant. The model achieved a best accuracy of 97% (Hassan et al., 2022).

Lombardi et al. looked at how XAI could predict cognitive decline and AD, generally known as Alzheimer's disease. They performed a three-way categorization between healthy control subjects, persons with cognitive impairment, and patients with AD using various cognitive indices. They provided a solid framework for analyzing the explainability variance linked to SHAP values. They demonstrated that the impact of each indicator on a patient's cognitive status can be accurately described by SHAP values. They claimed that a longitudinal analysis of SHAP values could offer useful insights into how AD develops. To forecast the sickness, they employed the ADNIMERGE R package and random forest classifiers (Lombardi et al., 2022). Lamy et al. interpreted breast cancer tomography images using AI methods and adapted them to Case-Based Reasoning technology. They classified publicly available datasets with mass size between 0–9 and benign-malignant. kNN, WkNN, and RBIA algorithms were used for testing and the test results facilitated interpretability by applying multidimensional visualization. As a result, the RBIA multidimensional visualization algorithm gave better results than the other two algorithms with an accuracy rate of 97% (Lamy et al., 2019). In their study, Karim et al. provided automatic decision-making with a deep neural network method that can be explained from chest radiography (CXR) images of Covid-19 symptoms. They used CXR images of 15,959 patients with Covid-19 cases as a dataset. They pre-processed CXR images with VGG19, ResNet-18, and DenseNet-161 algorithms and then highlighted the affected regions in XAI using GRAD CAM++ and layer-wise relevance propagation. As a result, explainable deep neural networks were able to identify 96.1% of Covid-19 cases in CXR images (Karim et al., 2020).

In their study, Chen et al. detected mammography images for early detection of breast cancer with AI methods and interpreted the data results with XAI methods. Using the Breast Cancer Causal XAI Diagnostic Model (BCCX-DM), they converted the mammography pictures into tabular data. The mammography pictures were categorized by the model as benign or malignant. They developed Causal-TabNet, which combines the causal graphs and the widely used tabular learning technique TabNet to enable reasoning in graphs and preserve feature connection. They found that their interpretable results were close to clinician interpretations (Chen et al., 2021). Smucny et al. used brain imaging AX-CPT data to identify frontoparietal activation associated with cognitive control in early psychosis using machine learning and deep learning algorithms. Naive Bayes, support vector machine, K Star, AdaBoost, J48 decision tree, and random forest machine learning techniques were used to assess AX-CPT images. They used the XAI model to analyze the deep learning results in order to make sense of them. Dorsolateral prefrontal cortex (DLPFC) was discovered by XAI to be the most predictive feature utilized by binary DL. The findings demonstrated that deep learning is more effective than superficial machine learning techniques in predicting symptomatic improvement (Smucny et al., 2021). Galazzo et al. used neuroimaging endophenotypes to estimate brain age using deep learning algorithms. They developed a lightweight DL architecture, simple full convolutional network (SFCN), based on VGGnet that takes whole brain 3D T1w images. The deep learning model was interpreted with a Pos-hoc explainable AI model (Boscolo Galazzo et al., 2022). In their study, Nayak et al. tried to early diagnose the monkeypox disease that emerged in early 2022 with AI methods. They classified chickenpox, measles, and healthy patients using the SqueezeNet deep learning model. The result of the deep learning model was visualized by applying LIME and the result could be interpreted. Monkeypox disease achieved 91.19% accuracy and a 92.55% F1 score with the algorithm used (Nayak et al., 2023). De Souza et al. used computer-aided Barrett's oesophagus and adenocarcinoma early detection applications with different deep neural networks to detect laryngeal cancer. Two different datasets consisting of six endoscopy images of the larynx were trained with four different convolutional neural networks (AlexNet, SqueezeNet, ResNet50, and VGG16). In practice, the location and number of features within the cancerous region were analyzed. Five different XAI methods were applied to interpret the results of the CNN architecture. The results of each model were interpreted using saliency (SAL), guided backpropagation (GBP), integrated gradients (IGR), input × gradients (IXG), and DeepLIFT (DLF) XAI. The results of the application showed the best agreement with the diagnoses of manual experts (de Souza et al., 2021).

5.6 CONCLUSIONS

The processing of medical pictures using XAI has advanced significantly because of this effort. With the aid of the created XAI algorithms, we can better comprehend the intricate details in medical photos. It demonstrates the enormous potential of XAI techniques for studying medical images. These techniques can aid in the early diagnosis of diseases and help us better understand the reasons of pathologies that have been found. They can also increase the accuracy of medical diagnoses. This study does, however, have certain shortcomings. For instance, it is important to remember that the descriptions obtained via XAI approaches are occasionally inaccurate and incomplete. To increase the precision and dependability of XAI approaches, additional investigation and validation studies are required. To sum up, XAI processing of medical pictures can be crucial to the processes of medical diagnosis and therapy. Medical professionals can make accurate and quick decisions by using XAI approaches to help them better grasp key information in photos.

REFERENCES

Al Hammadi, A. Y., Yeun, C. Y., Damiani, E., Yoo, P. D., Hu, J., Yeun, H. K., & Yim, M. S. (2021). Explainable artificial intelligence to evaluate industrial internal security using EEG signals in IoT framework. *Ad Hoc Networks*, *123*, 102641. https://doi.org/10.1016/J.ADHOC.2021.102641

Anand, A., Kadian, T., Shetty, M. K., & Gupta, A. (2022). Explainable AI decision model for ECG data of cardiac disorders. *Biomedical Signal Processing and Control*, *75*, 103584. https://doi.org/10.1016/J.BSPC.2022.103584

Banerjee, P., Banerjee, S., & Barnwal, R. P. (2022). Explaining deep-learning models using gradient-based localization for reliable tea-leaves classifications. *Proceedings of*

IEEE 2022 4th International Conference on Advances in Electronics, Computers and Communications, ICAECC 2022. https://doi.org/10.1109/ICAECC54045.2022.9716699

Boscolo Galazzo, I., Cruciani, F., Brusini, L., Salih, A., Radeva, P., Storti, S. F., & Menegaz, G. (2022). Explainable artificial intelligence for magnetic resonance imaging aging brainprints: Grounds and challenges. *IEEE Signal Processing Magazine, 39*(2), 99–116. https://doi.org/10.1109/MSP.2021.3126573

Chen, D., Zhao, H., He, J., Pan, Q., & Zhao, W. (2021). An causal XAI diagnostic model for breast cancer based on mammography reports. *Proceedings – 2021 IEEE International Conference on Bioinformatics and Biomedicine, BIBM 2021*, 3341–3349. https://doi.org/10.1109/BIBM52615.2021.9669648

de Souza, L. A., Mendel, R., Strasser, S., Ebigbo, A., Probst, A., Messmann, H., Papa, J. P., & Palm, C. (2021). Convolutional Neural Networks for the evaluation of cancer in Barrett's esophagus: Explainable AI to lighten up the black-box. *Computers in Biology and Medicine, 135*, 104578. https://doi.org/10.1016/J.COMPBIOMED.2021.104578

Fessler, J. (2010). Model-based image reconstruction for MRI. *IEEE Signal Processing Magazine, 27*(4), 81–89. https://doi.org/10.1109/MSP.2010.936726

Gøtzsche, P. C., & Jørgensen, K. J. (2013). Screening for breast cancer with mammography. *Cochrane Database of Systematic Reviews, 2013*(6). https://doi.org/10.1002/14651858.CD001877.PUB5/MEDIA/CDSR/CD001877/IMAGE_N/NCD001877-CMP-001-21.PNG

Hassan, M. R., Islam, M. F., Uddin, M. Z., Ghoshal, G., Hassan, M. M., Huda, S., & Fortino, G. (2022). Prostate cancer classification from ultrasound and MRI images using deep learning based Explainable Artificial Intelligence. *Future Generation Computer Systems, 127*, 462–472. https://doi.org/10.1016/J.FUTURE.2021.09.030

Hu, Q., Gois, F. N. B., Costa, R., Zhang, L., Yin, L., Magaia, N., & de Albuquerque, V. H. C. (2022). Explainable artificial intelligence-based edge fuzzy images for COVID-19 detection and identification. *Applied Soft Computing, 123*, 108966. https://doi.org/10.1016/J.ASOC.2022.108966

Jahmunah, V., Ng, E. Y. K., Tan, R. S., Oh, S. L., & Acharya, U. R. (2022). Explainable detection of myocardial infarction using deep learning models with Grad-CAM technique on ECG signals. *Computers in Biology and Medicine, 146*, 105550. https://doi.org/10.1016/J.COMPBIOMED.2022.105550

Jouis, G., Mouchère, H., Picarougne, F., & Hardouin, A. (2021). Anchors vs attention: Comparing XAI on a real-life use case. Lecture Notes in Computer Science (Including Subseries Lecture Notes in Artificial Intelligence and Lecture Notes in Bioinformatics), *12663 LNCS*, 219–227. https://doi.org/10.1007/978-3-030-68796-0_16/TABLES/5

Jung, Y. J., Han, S. H., & Choi, H. J. (2021). Explaining CNN and RNN using selective layer-wise relevance propagation. *IEEE Access, 9*, 18670–18681. https://doi.org/10.1109/ACCESS.2021.3051171

Karim, M. R., Dohmen, T., Cochez, M., Beyan, O., Rebholz-Schuhmann, D., & Decker, S. (2020). DeepCOVIDExplainer: Explainable COVID-19 diagnosis from chest X-ray images. Proceedings – 2020 IEEE International Conference on Bioinformatics and Biomedicine, BIBM 2020, 1034–1037. https://doi.org/10.1109/BIBM49941.2020.9313304

Lamy, J. B., Sekar, B., Guezennec, G., Bouaud, J., & Séroussi, B. (2019). Explainable artificial intelligence for breast cancer: A visual case-based reasoning approach. *Artificial Intelligence in Medicine, 94*, 42–53. https://doi.org/10.1016/J.ARTMED.2019.01.001

Lombardi, A., Diacono, D., Amoroso, N., Biecek, P., Monaco, A., Bellantuono, L., Pantaleo, E., Logroscino, G., De Blasi, R., Tangaro, S., & Bellotti, R. (2022). A robust framework to investigate the reliability and stability of explainable artificial intelligence markers of Mild Cognitive Impairment and Alzheimer's Disease. *Brain Informatics*, *9*(1), 1–17. https://doi.org/10.1186/S40708-022-00165-5/FIGURES/7

Maisey, M. N. (2005). Positron Emission Tomography in Clinical Medicine. U: Bailey DL, Townsend DW, Valk PE, Maisey MN, urednici. Positron emission tomography: Basic sciences. https://doi.org/10.1007/B136169

Nagaraj, P., Muneeswaran, V., Dharanidharan, A., Balananthanan, K., Arunkumar, M., & Rajkumar, C. (2022). A prediction and recommendation system for Diabetes Mellitus using XAI-based Lime Explainer. International Conference on Sustainable Computing and Data Communication Systems, ICSCDS 2022 – Proceedings, 1472–1478. https://doi.org/10.1109/ICSCDS53736.2022.9760847

Nayak, T., Chadaga, K., Sampathila, N., Mayrose, H., Bairy, G. M., Prabhu, S., Katta, S. S., & Umakanth, S. (2023). Detection of Monkeypox from skin lesion images using deep learning networks and explainable artificial intelligence, *Applied Mathematics in Science and Engineering*, *31*(1), 2225698. https://doi.org/10.1080/27690911.2023.2225698

Örnek, A. H., & Ceylan, M. (2020, October). Explainable features in classification of neonatal thermograms. In *2020 28th Signal Processing and Communications Applications Conference (SIU)* (pp. 1–4). IEEE. https://doi.org/10.1109/SIU49456.2020.9302311

Roshan, K., & Zafar, A. (2022). Using Kernel SHAP XAI method to optimize the network anomaly detection model. Proceedings of the 2022 9th International Conference on Computing for Sustainable Global Development, INDIACom *2022*, 74–80. https://doi.org/10.23919/INDIACOM54597.2022.9763241

Smucny, J., Davidson, I., & Carter, C. S. (2021). Comparing machine and deep learning-based algorithms for prediction of clinical improvement in psychosis with functional magnetic resonance imaging. *Human Brain Mapping*, *42*(4), 1197–1205. https://doi.org/10.1002/HBM.25286

Sudar, K. M., Nagaraj, P., Nithisaa, S., Aishwarya, R., Aakash, M., & Lakshmi, S. I. (2022). Alzheimer's disease analysis using Explainable Artificial Intelligence (XAI). International Conference on Sustainable Computing and Data Communication Systems, ICSCDS 2022 – Proceedings, 419–423. https://doi.org/10.1109/ICSCDS53736.2022.9760858

Tjoa, E., & Guan, C. (2021). A survey on explainable artificial intelligence (XAI): Toward medical XAI. *IEEE Transactions on Neural Networks and Learning Systems*, *32*(11), 4793–4813. https://doi.org/10.1109/TNNLS.2020.3027314

van der Velden, B. H. M., Kuijf, H. J., Gilhuijs, K. G. A., & Viergever, M. A. (2022). Explainable artificial intelligence (XAI) in deep learning-based medical image analysis. *Medical Image Analysis*, *79*, 102470. https://doi.org/10.1016/J.MEDIA.2022.102470

Withers, P. J., Bouman, C., Carmignato, S., Cnudde, V., Grimaldi, D., Hagen, C. K., Maire, E., Manley, M., Du Plessis, A., & Stock, S. R. (2021). X-ray computed tomography. *Nature Reviews Methods Primers*, *1*(1), 1–21. https://doi.org/10.1038/s43586-021-00015-4

6 Current Progress and Open Research Challenges for XAI in Deep Learning across Medical Imaging

Satish Kumar

6.1 INTRODUCTION

In last few years, there have been notable advancements in the field of medical imaging, largely driven by the adoption of deep learning-based algorithms (Litjens et al., 2017). These algorithms have played a pivotal role in revolutionising the medical imaging with enhancing the efficiency and accuracy of diagnosis. However, the major challenge that prevents the widespread use of deep leaning algorithms in clinical settings is their lack of interpretability and transparency. To tackle this obstacle, researchers have embraced explainable AI (XAI) with regard to provide clear and comprehensible solutions. XAI provides techniques and models that reveal the underlying processes of deep learning algorithms, making them more transparent and understandable (Holzinger et al., 2017, Leopold et al., 2020). The utilization of learning-based methods and models in medical imaging tasks has developed impressive outcomes. For instance, lung cancer detection, Alzheimer's classification, and diagnosing diseases have seen the implementation of deep learning algorithms (Sarraf & Aghdam 2018, Liu et al., 2020, Burlina et al., 2021, Jo T et al., 2019). However, despite these achievements, there are still challenges and limitations in implementing AI-based methods within settings (Rajkomar et al., 2019, Sengupta et al., 2020).

The main goal of this chapter is to explore current progress and figure out open research challenges in applying XAI to deep learning for medical imaging. Further, we aim to expand the boundaries of this exciting field of research and set the way for more advancements in implementing XAI into clinical practice (Holzinger et al., 2019). AI limitations can be caused by a variety of factors, including deep learning models' black-box nature and the associated cost of computation. Although deep learning-based models are mostly based on statistical rules, they lack explicit

DOI: 10.1201/9781003426073-6

representation of specific tasks knowledge and pose interpretability challenges. In contrast, fundamental AI techniques such as decision trees and linear regression somehow provide automatic explanations. These basic methods allow for visualising decision boundaries and understanding classification in lower dimensions, but they lack the computational operation required for higher dimensions that involve 3D and the majority of 2D medical images. These limitations of basic methods has inspired the development of more applications such as in autonomous driving and finance, where reliability and explainability are essential in establishing some kind of certainty with end-users (Smith, 2019).

In the biomedical field, the adoption of AI systems for diagnosis to establish certainty among physicians, regulators, and patients is important. Transparent and explainable systems are necessary to get around the underlying rationale behind their decision-making process, enabling stakeholders to easily engage with them (Rudin, 2019, Lipton,2018). The mostly commonly used deep learning-based black-box algorithms, which lack transparency, experience difficulties due to potential regulations, namely the European General Data Protection Regulation (GDPR) (Singh, et al., 2020). These regulations now require traceability of decisions, limiting the application of opaque models (Goodman & Flaxman, 2017). In order to effectively support medical professionals, AI systems should incorporate a certain level of explainability, enabling human experts to understand the decision-making process and exercise their judgment. It is important to understand that, under some circumstances, even humans may not always be able or ready to explain their choices. However, explainability plays a crucial role in ensuring fair, safe, ethical, and reliable AI utilization, thus enabling its practical implementation (Lipton, 2018). By demystifying AI and revealing the factors considered during the decision-making process, trust among end-users can be fostered, especially among non-deep learning end users like medical experts who benefit from understanding the specific features used in decision-making (Caruana et al., 2015). The phrases "explainability" and "interpretability" are sometimes used interchangeably, although a study proposes a clear separation between these notions (Lipton, 2018). In this study, "interpretation" was stated as the process of converting abstract concepts, such as output classes, to concrete examples in the domain. In contrast, "explanation" refers to a collection of domain-specific information, such as pixels in images, that helps the model draw a conclusion. The concept of "uncertainty" is intimately tied to the decision-making process of a model.

In situations where ambiguity exists, deep learning classifiers often struggle to convey uncertainty and tend to provide the class with the highest probability, even when the margin is slim. Recent research has explored the integration of uncertainty analysis with explainability to address cases where the model is unsure, making them more suitable for non-deep learning users. The lack of transparency in deep learning models arises from the inability to directly understand or interpret the weights of neurons. Another study (Smith, 2019) found that the size and selectivity of activations, as well as their impact on network decisions, are insufficient to assess the relevance of a neuron for a given task. A complete examination of terminologies, concepts, and application cases linked to XAI can be found in another reference (Bas and Velden, 2022).

XAI approaches can provide a clear and transparent understanding of deep learning algorithms' decision-making processes. By providing explainable results, XAI can help build trust in these algorithms and improve their clinical impact.

This chapter provides an in-depth examination of works concentrating on the explainability of deep learning-based models in the context of medical imaging. In Section 2, an extensive taxonomy of explainability techniques is briefly described. Section 3 discusses recent advances in XAI in medical image analysis. Furthermore, Section 4 provides the details of various applications of XAI. The challenges of XAI are discussed in Section 5. Finally, Section 6 we examine the evolution of explainable deep learning models in medical image analysis, present trends, and potential future advances.

6.1.1 Objectives

The main goal of this chapter is to discuss the XAI boundaries in deep learning for medical imaging and to figure out the most recent advances and open research problems. The specific contributions of this chapter are as follows:

- To provide an outline for the current state-of-the-art XAI methods for deep learning in medical imaging.
- To discuss the benefits and limitations of XAI in medical imaging, including its impact on clinical practice and patient outcomes.
- To identify open research issues for using XAI in deep learning for medical imaging, such as bias reduction, improved interpretability, and performance optimization.
- To highlight future research directions in XAI for medical imaging, such as innovative techniques, data-sharing efforts, and multidisciplinary collaborations.

6.2 EXPLAINABILITY AI APPROACHES

In this study, various taxonomies have been suggested to classify the wide array of XAI methods. It is essential to point out that these classification strategies are not rigorous and might vary significantly depending on the methods' individual properties. As a result, the methods might be classified into many overlapping or non-overlapping classes at the same time. For a more comprehensive analysis of these taxonomies, refer to Bas and Velden (2022) and Stano et al. (2020). Additionally, Figure 6.1 presents a helpful flowchart summarising the taxonomies.

6.2.1 Post-hoc Explainability Approaches

Post-hoc explainability approaches focus on explaining the decisions of AI models after they have made predictions. These methods do not require changes in the base model architecture while training. Instead, they analyse the model's outputs and internal representations to generate explanations. Saliency maps, which emphasise the significant aspects in the input data, and gradient-based techniques, which

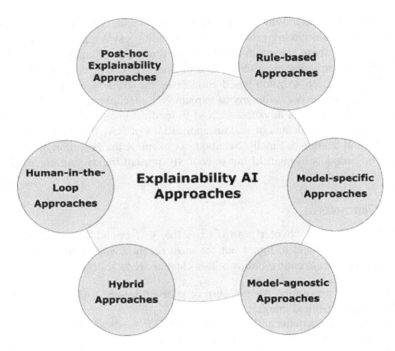

FIGURE 6.1   XAI approaches.

analyse the model's gradients with respect to the input, are two examples of post-hoc explainability approaches.

6.2.2   RULE-BASED APPROACHES

Rule-based techniques seek to explain things using logical rules or decision trees. These guidelines are simple for humans to understand and aid in the decision-making process of AI models. Rule-based techniques sometimes entail extracting rules from the learned model or developing simplified rule-based models that match the original model's behavior.

6.2.3   MODEL-SPECIFIC APPROACHES

Model-specific approaches are tailored to specific types of AI models, taking advantage of their unique characteristics to generate explanations. For example, in deep learning-based models and methods, deconvolution networks can help visualise the activations of different layers, providing insights into the hierarchical representation of features. Similarly, in ensemble models, explanations can be derived from the contributions of individual models within the ensemble.

6.2.4 MODEL-AGNOSTIC APPROACHES

Model-agnostic techniques are independent of the base AI model and can be used for any model. They want to provide details that can be applied to several AI-based models. Model-agnostic approaches that include permutation importance or Shapley values frequently involve modifying the data that goes into the model and evaluating the changes in the model's predictions.

6.2.5 HYBRID APPROACHES

Hybrid methods employ a combination of explanation techniques to generate precise summary. These methods take advantage of the strengths of approaches while aiming to mitigate their limitations. For instance a hybrid approach could integrate rule-based explanations, with hoc techniques offering both easily understandable rules and detailed feature importance analysis.

6.2.6 HUMAN-IN-THE-LOOP APPROACHES

Human-in-the-loop strategies bringing real-world experts to assist and demonstrate objects. These approaches developed interactive and iterative mechanisms that enable humans to investigate and refine the explanations given using AI models. User feedback and iterative improvement of these explanations based on preferences and domain knowledge are frequently employed in human-in-the-loop approaches.

Each of these classifications are not mutually exclusive, and many explainability approaches can include or overlap components from multiple categories. The best explainability approach determines the factors such as the AI model used, the intended audience, the nature of the application, and the desired level of interpretability.

6.3 RECENT ADVANCES OF EXPLAINABILITY AI IN MEDICAL IMAGE ANALYSIS

Medical image analysis based on AI is mostly used in disease diagnosis, treatment planning, and monitoring. Due to the increasing use of deep learning algorithms in this domain, there is an extensive demand for XAI methodologies to improve the interpretability and reliability of AI models. In this section we will cover recent advances in XAI techniques applied to medical image analysis and briefly talk about their contributions to improving diagnostic accuracy, treatment decision-making, and overall patient care.

6.3.1 INTERPRETABILITY THROUGH ATTENTION MECHANISMS

In medical image processing, attention mechanisms have been identified as highly efficient interpretability tools. Recent research studies have shown that attention-based models, such as self-attention networks and spatial transformer networks, can be employed to highlight relevant regions or characteristics in medical images. These attention approaches offer valuable insight into the decision-making process of AI models and assist radiologists in understanding and validating the model's predictions (Zhang, et al., 2020).

6.3.2 MODEL-AGNOSTIC EXPLANATIONS

In the past few years, model-agnostic explainability approaches have received a significant amount of attention, due to their ability to deliver explanations for AI models. Model-agnostic approaches that include permutation importance or Shapley values frequently involve modifying the data that goes into the model and evaluating the changes in the model's predictions. These methodologies provide medical practitioners with insight into the decision-making processes of sophisticated AI models, regardless of their underlying architecture (Lundberg & Lee, 2017).

6.3.3 RULE-BASED EXPLANATIONS

Rule-based explanations have shown major development in biomedical image analysis, providing transparent decision rules based on AI models. These standards present explicit criteria for decision-making, permitting most clinicians to gain insight into the acute reasoning behind a prediction. Research studies have proposed methods for extracting decision rules from deep learning models or to generate simplified rule-based models that represent the behaviour of the original model. Rule-based explanations assist researchers to build trust in AI systems by providing transparent and comprehensible insights into the model's decision-making system (Zhang et al., 2020).

6.3.4 INTERACTIVE AND HUMAN-IN-THE-LOOP APPROACHES

The dynamic and human-in-the-loop approaches have gained popularity due to the fact that they incorporate active participation in medical professionals for explainability. These methods allow clinicians to provide feedback, improve explanations, and incorporate domain knowledge into the AI models. In relation to human experts in the interpretability process, these methods improve the accuracy and significance of the explanations, resulting in a greater degree of decision-making and collaboration between clinicians and AI system (Caruana et al., 2015).

6.3.5 INTEGRATION OF CLINICAL DATA AND PRIOR KNOWLEDGE

The most recent advances in XAI tend to focus on incorporating clinical data and prior knowledge towards the interpretation process. XAI models can represent customised explanations and insights tailored to individual patients by utilising patient-specific clinical data, including electronic health records or genomic information. Combining prior medical knowledge into the interpretability process improves clinical accuracy and promotes evidence-based decision-making (Rajkomar et al., 2018).

Recent advances in XAI have the potential to transform medical image analysis by allowing clinicians to better comprehend and trust AI models. These methods are designed to bridge the gap between the black-box nature of deep learning models and the need for transparency and accountability in medical decision-making with interpretable explanations. However, issues such as scalability, robustness, and validation of all of these techniques in clinical settings must be addressed additionally to ensure their successful incorporation into routine clinical practise.

6.4 APPLICATIONS OF XAI

This section covers the explainability applications in medical imaging and divides them into two parts: pre-existing attribution-based approaches and unique methodologies. We address the techniques and implications for medical imaging applications based on the type of explainability approach used.

6.4.1 ATTRIBUTION BASED

The wide majority of medical imaging research now focuses on the interpretability of deep learning algorithms by utilising attribution-based approaches. These methods allow researchers to train appropriate neural network topologies without taking on the task of making them explainable. This flexibility enables the use of pre-existing deep learning models and also enables the development of novel architectures for optimum performance on particular tasks.

The former approach makes implementation easier and enables approaches such as transfer learning, but the latter can focus on specific data and reduce overfitting by utilising fewer parameters. Both methods are useful for medical imaging datasets, which are often smaller than computer vision benchmarks like ImageNet. Researchers can check whether the model is learning relevant characteristics or relying on false ones by analysing attributions in post-model analysis and allowing tweaks to the model architecture and hyper-parameters to produce better outcomes on test data. As a result, the model's potential for real-world applications grows. In this section, we look at current research that uses attribution methods in medical imaging modalities such as brain magnetic resonance imaging (MRI), retinal imaging, breast imaging, skin imaging, computerized tomography (CT) scans, and chest X-rays.

6.4.1.1 Breast Imaging

Using Integrated Gradients (IG) and SmoothGrad approaches, the characteristics of a Convolutional Neural Network (CNN) utilised for categorising oestrogen receptor status from breast MRI were visualised (Santos et al., 2021). According to the results of the investigation, the model successfully learned relevant features in both the spatial and dynamic domains, with separate contributions from each. The visualisations, however, revealed the presence of several irrelevant characteristics, which were attributed to pre-processing artefacts. These findings spurred changes to the pre-processing and training methods. Two other CNNs, AlexNet and GoogleNet, were used in a related work concentrating on breast mass categorisation from mammograms (Nakahara et al., 2020). The image properties learned by these CNN models were visualised using saliency maps. Interestingly, both models demonstrated the ability to learn the boundaries of the mass, which is an important clinical requirement. They were also sensitive to the contextual information provided in the photos.

6.4.1.2   X-ray Imaging

A method named GSInquire was utilised in a recent work focusing on the identification of COVID-19 from chest X-ray images to create heatmaps for confirming the characteristics learned by the proposed COVID-net model (Khan et al., 2022). In terms of the newly presented measures – Impact Score and Impact Coverage – GSInquire, an attribution method, outperformed existing methods such as SHAP and Expected Gradients. The Impact Score was defined as the percentage of features that significantly influenced the model's judgement or confidence, whereas Impact Coverage was the coverage of adversarially influenced components in the input.

Another study looked at the ambiguity and interpretability of chest X-rays for COVID-19 detection (Kassani et al., 2021). The study used a variety of strategies to generate heatmaps for the trained model's sample inputs, including saliency maps, Guided Grad-CAM, GBP, and Class Activation Maps (CAM). These visualisation methods provided useful insights into the sections of the image that contributed the most to the model's predictions and aided in comprehending the decision-making process.

6.4.1.3   CT Imaging

A unique attribution method inspired by DeepDreams was developed in a recent work focusing on explaining tumour segmentation from liver CT scans (Mordvintsev et al., 2020). This method, like other attribution methods, used the DeepDreams idea, an image generating algorithm, and could be applied to black-box neural networks. The strategy maximised the activation of the target neuron through gradient ascending by doing sensitivity analysis, essentially identifying the steepest slope of the function.

Attribution-based methods were widely utilised as early approaches to visualising neural networks, progressing from simple class activation maps and gradient-based methods to complex techniques like as DeepSHAP (Hua et al., 2015). These enhanced visualisation methods indicated that the models learned meaningful elements in the vast majority of situations. Instances of false features were thoroughly investigated, noted to readers, and resulted in changes to the model training techniques. The use of smaller and task-specific models, in conjunction with customised forms of attribution methods, can improve the detection of relevant elements even more. Researchers can acquire better insights into the decision-making processes of deep learning models used in medical image analysis by exploiting these methodologies.

6.4.2   Attention Based

Deep learning has seen a surge in the use and utility of attention techniques. These methods, inspired by human attention, allow models to focus on different regions of an image or data source for analysis. Vaswani et al. (2017) go into greater detail about attention mechanisms in neural networks. The study by Zhang et al. (2020) provides an example of attention in medical diagnosis. We investigate how attention-based algorithms can be used as explainable deep learning tools for medical picture analysis in this research.

To develop a direct mapping between medical images and accompanying diagnostic reports, a network dubbed MDNet was proposed (Zhang et al., 2020). Attention mechanisms were employed to visualise the detection process by employing image and language models. The language model discovered prominent and discriminative elements using the attention mechanism to learn the mapping between images and diagnostic reports. This pioneering research was the first to use attention mechanisms to extract useful information from medical image databases. The use of attention-based approaches in medical image analysis improves interpretability and transparency, giving doctors and researchers useful insights.

6.4.3 TEXTUAL JUSTIFICATION

A justification model was created to improve the explainability of deep learning models, allowing the model to provide reasoning in the form of words or phrases that directly connect with both expert and general users. In the context of breast mass classification, a justification model was created (Lee et. Al., 2019) that used inputs from a classifier's visual characteristics as well as prediction embeddings. To clarify categorization decisions, our model provided diagnostic words and visual heatmaps.

6.5 CHALLENGES

While explainability AI methods have shown promising results in improving the interpretability of deep learning models in medical image analysis, various challenges must be overcome before they can be successfully implemented and widely adopted in clinical practice. These difficulties include the following.

6.5.1 COMPLEXITY OF DEEP LEARNING MODELS

Deep learning methods and models used in medical image analysis are often complex and have a large number of parameters. Understanding the internal workings of these models and providing meaningful explanations for their decisions can be challenging. Developing explainability methods that can effectively handle the complexity of deep learning architectures and provide clear and intuitive explanations is a key challenge (Litjens et al., 2017, Lundervold&Lundervold, 2019).

6.5.2 INTERPRETABILITY VERSUS PERFORMANCE TRADE-OFF

There is usually a trade-off between the interpretability of models and performance. Some highly interpretable models may sacrifice performance, while complex models with high accuracy may lack interpretability. Striking the right balance between model performance and interpretability is crucial, as clinicians require both accurate predictions and understandable explanations to make informed decisions (Caruana et al., 2015, BarredoArrieta et al., 2020).

6.5.3 LACK OF STANDARD EVALUATION METRICS

The evaluation of explainability approaches in medical image analysis is currently in its early phases, and also standardised. Currently, different studies employ various evaluation criteria, making it challenging to compare and benchmark different approaches. Developing robust evaluation metrics that assess the quality, interpretability, and clinical relevance of the explanations is necessary to ensure the validity and reliability of explainability AI methods (Doshi-Velez et al., 2017, Holzinger et al., 2018).

6.5.4 LIMITED AVAILABILITY OF LABELED DATA

Explainability AI methods often require annotated data for training and validation. However, collecting large-scale annotation-based data sets for medical imaging is difficult due to privacy concerns and the need for expert annotations. The limited availability of labeled data hampers the development and evaluation of explainability methods. Exploring strategies such as transfer learning, data augmentation, and active learning to address the issue of data scarcity is a significant topic of future research (Wang et al., 2020, Maier-Hein et al., 2018).

6.5.5 ETHICAL AND LEGAL IMPLICATIONS

The AI usage in healthcare raises ethical and legal considerations. XAI should adhere to patient privacy regulations and ensure that explanations are fair, unbiased, and transparent. Additionally, the responsibility and accountability of the decisions made by AI systems need to be clearly defined. Developing guidelines and standards for the ethical implementation of explainability AI in medical image analysis is essential to maintain patient trust and ensure responsible use (Obermeyer & Emanuel, 2016, Char et al., 2020).

6.5.6 INTEGRATION WITH CLINICAL WORKFLOW

For explainability AI methods to be effectively integrated into clinical practice, they need to seamlessly fit within the existing clinical workflow. Clinicians and healthcare professionals should be able to easily interpret and trust the explanations provided by AI systems without disrupting their decision-making processes. Developing user-friendly interfaces and tools that present explanations in a concise and actionable manner is critical for successful integration (Yadav et al., 2020, Chen et al., 2017).

Addressing these challenges will contribute to the wider adoption and practical implementation of explainability AI in medical image analysis. Future research should focus on developing advanced and robust explainability methods, establishing standardised evaluation frameworks, exploring strategies for handling limited labelled data, ensuring ethical and legal compliance, and designing user-centric interfaces for effective integration into clinical workflows.

6.6 DISCUSSION

The use of explainability AI approaches in medical image analysis has shown tremendous promise for improving the interpretability and transparency of deep learning algorithms. In this study, we investigated various explainability techniques and their applications across different medical imaging modalities. The findings emphasise the importance of explainability in the medical arena, where the ability to understand and interpret deep learning models' judgements is critical for obtaining physicians' and researchers' trust and approval. One key observation from our analysis is the prevalence of attribution-based methods in the literature. Grad-CAM, IG, and SmoothGrad have all proven to be useful tools for visualising the features gained by deep learning models. They provide insights into the regions of interest within the photos, which help the model make decisions. Researchers can examine the model's concentration on clinically relevant aspects and identify the presence of spurious or irrelevant features by scrutinising these visualizations. This enables adjustments to be made to the training methods, such as preprocessing techniques or model architectures, to enhance the model's performance on specific medical imaging tasks. Furthermore, the implementation of attention mechanisms in the analysis of medical images has shown promising results. These mechanisms, inspired by human attention, allow algorithms to selectively focus on the important locations of an image or data source. Researchers can acquire insights into the model's decision-making process and learn which sections of the input data contribute the most significantly to the model's predictions with visualising the attention maps. Attention-based methods have facilitated the development of explainable deep learning models that can directly communicate their reasoning to both expert and general users. This fosters user confidence and facilitates collaboration between AI systems and healthcare professionals.

There have been efforts to generate textual justifications for deep learning model decisions in medical image analysis along with visual explanations. Justification models provide diagnostic sentences and visual heatmaps in order to offer reasoning for the model's decisions while utilising the visual features of classifiers and prediction embedding. The accessibility of multimodal explanations, containing both textual and visual data, increases user confidence and enhances understanding of the model's decision-making process. Whereas significant progress has already been made in the field of XAI in medical image analysis, there are still a number of challenges and areas for future research. An example of this is that there is a demand for standardised evaluation metrics to assess the quality and reliability of explainability approaches. Additionally, there is currently no consensus in the medical sector on the best way to accurately evaluate the interpretability of deep learning models. The design and implementation of comprehensive assessment frameworks will aid in the comparison and benchmarking of various methodologies and contribute to the use of explainability AI in clinical practice. Furthermore, combining XAI techniques with other emerging technologies, including transfer learning, reinforcement learning, and generative models, has the potential to improve deep learning models' interpretability and generalisability in medical image analysis. Researchers can improve the explainability of complex models as well as tackle issues related to limited labelled data and domain adaptation by combining these approaches.

Finally, XAI has been proven to be an important area of research in medical image analysis, permitting clinicians and researchers to support decisions based on deep learning methods and models. Attribution-based methods, justification models, and attention mechanisms have contributed valuable information towards the model's decision-making procedure and assisted the interaction between human endusers and AI systems. In the future, studies will be focused on standardising evaluation metrics, figuring out unique integration strategies, and assigning the legal and ethical implications of XAI in medical practice. Continued research and the use of XAI in medical image analysis have the potential to lead to better patient care, clinical decision-making, and overall field advancement.

REFERENCES

Adadi, A., & Berrada, M. (2018). Peeking inside the black-box: A survey on explainable artificial intelligence (XAI). *IEEE Access*, 6, 52138–52160.

BarredoArrieta, A., Díaz-Rodríguez, N., Del Ser, J., Bennetot, A., Tabik, S., Barbado, A., … Herrera, F. (2020). Explainable artificial intelligence (XAI): Concepts, taxonomies, opportunities and challenges toward responsible AI. *Information Fusion*, 58, 82–115.

Burlina, P. M., Joshi, N., Pekala, M., Pacheco, K. D., Freund, D. E., & Bressler, N. M. (2021). Automated grading of age-related macular degeneration from color fundus images using deep convolutional neural networks. *JAMA Ophthalmology*, 139(7), 772–779.

Caruana, R., Lou, Y., Gehrke, J., Koch, P., Sturm, M., & Elhadad, N. (2015). Intelligible models for healthcare: Predicting pneumonia risk and hospital 30-day readmission. In *Proceedings of the 21th ACM SIGKDD International Conference on Knowledge Discovery and Data Mining*, Sydney NSW Australia, August 10–13 (pp. 1721–1730).

Char, D. S., Shah, N. H., Magnus, D. (2020). Implementing machine learning in health care—addressing ethical challenges. *New England Journal of Medicine*, 382, 1862–1864.

Chen, J. H., Asch, S. M. (2017). Machine learning and prediction in medicine—beyond the peak of inflated expectations. *New England Journal of Medicine*, 376, 2507–2509.

Doshi-Velez, F., Kim, B. (2017). Towards a rigorous science of interpretable machine learning. *arXiv preprint*, arXiv2017, arXiv:1702.08608.

Goodman, B., & Flaxman, S. (2017). European Union regulations on algorithmic decision-making and a "right to explanation". *AI Magazine*, 38(3), 50–57.

Holzinger, A., Biemann, C., Pattichis, C. S., & Kell, D. B. (2017). What do we need to build explainable AI systems for the medical domain? arXiv Preprint arXiv:1712.09923.

Holzinger, A., Langs, G., Denk, H., Zatloukal, K., & Müller, H. (2018). Causability and explainability of artificial intelligence in medicine. *Wiley Interdisciplinary Reviews: Data Mining and Knowledge Discovery*, 8(4), e1280.

Holzinger, A., Langs, G., Denk, H., Zatloukal, K., & Müller, H. (2019). Causability and explainability of artificial intelligence in medicine. *Wiley Interdisciplinary Reviews: Data Mining and Knowledge Discovery*, 9(4), e1312.

Holzinger, A., Plass, M., Holzinger, K., Crişan, G. C., Pintea, C. M., & Palade, V. (2017). Towards interpretable machine learning by extracting visual features from deep neural networks. In *International Conference on Information Technology in Bio- and Medical Informatics* (pp. 318–327). Springer.

Hua, K. L., Hsu, C. H., Hidayati, S. C., Cheng, W. H., & Chen, Y. J. (2015). Computer-aided classification of lung nodules on computed tomography images via deep learning technique. *OncoTargets and Therapy*, 11, 8.

Jo, T., Nho, K., & Saykin, A. J. (2019). Deep learning in Alzheimer's disease: Diagnostic classification and prognostic prediction using neuroimaging data. *Frontiers in Aging Neuroscience*, 11, 220.

Kassani, S. H., Kassasbeh, M., & Taher, F. (2021). COVID-XNet: A framework for efficient COVID-19 detection from chest X-ray images using neural networks and local graph kernels. *Computers in Biology and Medicine*, 134, 104505.

Khan, A. I., Hasan, S. M., Afridi, M. J., Rehman, A., Hussain, M., & Hassan, A. (2022). GSInquire: Gradient-based Saliency Inquire for Explainable AI in COVID-19 detection. *IEEE Journal of Biomedical and Health Informatics*, 26(2), 638–648.

Lee, H., Kim, S. T., Ro, Y. M. (2019). Generation of multimodal justification using visual word constraint model for explainable computer-aided diagnosis. In *Interpretability of Machine Intelligence in Medical Image Computing and Multimodal Learning for Clinical Decision Support*. Springer, Cham, pp. 21–29.

Leopold, H., Singh, A., Sengupta, S., Zelek, J., Lakshminarayanan, V. (2020). Recent Advances in Deep Learning Applications for Retinal Diagnosis using OCT. In State of the Art in Neural Networks, A. S. El-Baz (ed.). Elsevier, NY.

Lipton, Z. C. (2018). The mythos of model interpretability. In *Proceedings of the 2018 ICML Workshop on Human Interpretability in Machine Learning* (pp. 10–13).

Litjens, G., Kooi, T., Bejnordi, B. E., Setio, A. A., Ciompi, F., Ghafoorian, M., & Sanchez, C. I. (2017). A survey on deep learning in medical image analysis. *Medical Image Analysis*, 42, 60–88.

Liu, Y., Ji, X., Zhu, L., Li, W., & Chen, H. (2020). Deep learning in medical ultrasound analysis: A review. *Engineering*, 6(3), 291–307.

Lundervold, A. S., & Lundervold, A. (2019). An overview of deep learning in medical imaging focusing on MRI. *Zeitschrift für Medizinische Physik*, 29(2), 102–127.

Lundberg, S.M., & Lee, S. (2017). A unified approach to interpreting model predictions. In *Proceedings of the 31st Conference on Neural Information Processing Systems* (pp. 1–10), Long Beach, CA, USA, Dec. 4–9, 2017.

Maier-Hein, L., Eisenmann, M., Reinke, A., Onken, M., Scholz, P., & Meinzer, H. P. (2018). Why rankings of biomedical image analysis competitions should be interpreted with care. *Nature Communications*, 9(1), 5217.

Mordvintsev, A., Olah, C., & Tyka, M. (2015). Inceptionism: Going deeper into neural networks. Google AI Blog, https://ai.googleblog.com/2015/06/inceptionism-going-deeper-into-neural.html, accessed: 23 May 2020.

Nakahara, A., Nishimura, K., Uchiyama, Y., Nishio, M., & Sugano, S. (2020). Convolutional neural network models for breast mass classification using mammograms: A comparative study. *Radiological Physics and Technology*, 13(3), 285–292.

Obermeyer, Z., & Emanuel, E. J. (2016). Predicting the future—big data, machine learning, and clinical medicine. *New England Journal of Medicine*, 375(13), 1216–1219.

Rajkomar, A., Dean, J., & Kohane, I. (2019). Machine learning in medicine. *New England Journal of Medicine*, 380(14), 1347–1358.

Rajkomar, A., Oren, E., Chen, K., Dai, A. M., Hajaj, N., Hardt, M., Liu, P. J., Liu, X., Marcus, J., Sun, M., Sundberg, P., Yee, H., Zhang, K., Zhang, Y., Flores, G., Duggan, G. E., Irvine, J., Le, Q. V., Litsch, K., Mossin, A., Tansuwan, J., Wang, D., Wexler, J., Wilson, J., Ludwig, D., Volchenboum, S. L., Chou, K., Pearson, M., Madabushi, S., Shah, N. H.,

Butte, A. J., Howell, M. D., Cui, C., Corrado, G. S., & Dean, J. (2018). Scalable and accurate deep learning with electronic health records. *NPJ Digital Medicine*, 1(1), 1–10.

Ribeiro, M. T., Sameer and Guestrin, Carlos (2016). "Why should I trust you?": Explaining the predictions of any classifier. In *Proceedings of the 22nd ACM SIGKDD International Conference on Knowledge Discovery and Data Mining, USA* (pp. 1135–1144).

Rudin, C. (2019). Stop explaining black box machine learning models for high stakes decisions and use interpretable models instead. *Nature Machine Intelligence*, 1(5), 206–215.

Santos, D. D. S., Castro, W. L., Papa, J. P., & Costa, A. F. (2021). Deep learning applied to the classification of breast MRI: A study on the estrogen receptor status. *Journal of Digital Imaging*, 34(2), 402–412.

Sarraf, O., & Aghdam, K. A. (2018). Deep AD: Alzheimer's disease classification via deep convolutional neural networks using MRI and fMRI. *BioRxiv*, 070441.

Sengupta, S., Singh, A., Leopold, H. A., Gulati, T., & Lakshminarayanan, V. (2020). Ophthalmic diagnosis using deep learning with fundus images–A critical review. *Artificial Intelligence in Medicine*, 102, 101758.

Singh, A., Sengupta, S., & Lakshminarayanan, V. (2020). Explainable deep learning models in medical image analysis. Journal of Imaging, 6.

Smith, C. (2019). Explainable AI: From black box to glass box. *International Journal of Online and Biomedical Engineering*, 15(6), 146–157.

Stano, M., Benesova, W., & Martak, L.S. (2020). Explainable 3D convolutional neural network using GMM encoding. In *Proceedings of Twelfth International Conference on Machine Vision (ICMV 2019)*. Proceedings SPIE, 11433, p. 114331U.

Vaswani, A., Shazeer, N., Parmar, N., Uszkoreit, J., Jones, L., Gomez, A. N., ... & Polosukhin, I. (2017). Attention is all you need. *Advances in Neural Information Processing Systems,* 5998–6008.

Wang, X., Peng, Y., Lu, L., Lu, Z., Bagheri, M., & Summers, R. M. (2017). Chestx-ray8: Hospital-scale chest x-ray database and benchmarks on weakly-supervised classification and localization of common thorax diseases. In *Proceedings of the IEEE Conference on Computer Vision and Pattern Recognition (CVPR)*, Honolulu, HI, USA (pp. 3462–3471).

Yadav, S., Singh, P., Reddy, C. K., Ramakrishnan, G., & Babu, R. V. (2020). Interpretable deep learning models for medical image classification. In *2020 IEEE 17th India Council International Conference (INDICON), India* (pp. 1–6).

Zhang, H., Danaee, P., & Liu, J. (2020). MDNet: A unified diagnostic network for joint diseases from MRI images and clinical reports. *Journal of Biomedical Informatics*, 109, 103527.

Zhang, J., Gorriz, J.M., Dong, Z. (2020). Deep learning in medical image analysis: Recent advances and future trends. *European Journal of Radiology*, 122, 108770.

Wu, M., Hughes, M., Parbhoo, S., Zazzi, M., Roth, V., & Doshi-Velez, F. (2018). Beyond sparsity: Tree regularization of deep models for interpretability. In *Proceedings of the AAAI Conference on Artificial Intelligence*, 34(7), 10937–10944.

7 Towards Trustworthy and Reliable AI
The Next Frontier

Shalom Akhai

7.1 INTRODUCTION

Artificial intelligence (AI) has changed everything from healthcare diagnostics to recommendation system algorithms (Kumar et al. 2022). However, AI technologies, especially deep learning systems, are complicated, raising transparency, accountability, and ethical concerns (Santhoshkumar et al. 2023). Natural language processing, computer vision, predictive analytics, and robotics are all part of AI. More AI applications are causing need for explainable AI (Soori et al. 2023). This is a technological and social need (Hassija et al. 2023). Explainable AI systems explain their judgments, building confidence by demystifying AI's mystery (Theis et al. 2023). Explainable AI systems protect algorithms against biases and mistakes and allows for their correction. AI systems must be transparent to provide accountability and justice under ethical and legal norms (Albahri et al. 2023). AI interpretability is difficult beyond technical issues. One of the biggest issues is balancing model interpretability with performance. Scalability and complexity in modern AI models are major challenges (Embarak 2023). Human variables and cognitive biases complicate AI interpretability. This cognitive gap must be bridged to make AI system explanations accessible and understandable to varied audiences (Bertrand et al. 2022).

Interpretability in AI can be achieved in techniques such as decision trees, rule-based systems and linear models. But more advanced AI techniques including neural networks and hybrid models should be supported with post-hoc explanation methods. In this context, LIME and SHAP are widely used post-hoc explanation methods(Jin et al. 2022; Markus et al. 2021; Mehdiyev & Fettke 2021; Saleem et al. 2022; Srinivasu et al. 2022; Bharati et al. 2023). This chapter examines AI interpretability's complex issues, tactics, and research. It also predicts future developments in machine learning, multidisciplinary cooperation, ethics, and regulation.

7.2 BACKGROUND

AI interpretability improves transparency, responsibility, and trust by explaining AI system judgments. Complexity, interpretability-performance trade-offs, and human biases are challenges. Model-based methodologies, post-hoc explanation methods,

DOI: 10.1201/9781003426073-7

hybrid approaches, and ethical issues including openness, accountability, and justice in AI decision-making have been studied.

7.2.1 THE CONCEPT OF AI INTERPRETABILITY

AI interpretability is essential for technical progress and social integration, helping us comprehend AI (Reyes et al. 2020). It promotes openness, accountability, and confidence in AI system judgements as they evolve (Gilpin et al. 2018). The interpretability of AI enables explanation of its judgments and predictions, improving transparency and accountability. It also finds and fixes system flaws and biases, making it essential to AI development (Carvalho et al. 2019).

7.2.2 CHALLENGES OF AI INTERPRETABILITY

AI interpretability is complicated and needs careful consideration of many issues. Some models have higher interpretability but lower accuracy or efficiency. AI models, especially deep learning models, are scalable and sophisticated, making their decision-making processes harder to comprehend and evaluate. The size of these models makes communicating their results difficult (Papadimitroulas et al. 2021). User perception of sophisticated AI-generated judgments may also be influenced by human variables and cognitive biases. AI interpretability requires bridging the cognitive gap between users and AI systems to make explanations explicit and understandable to a varied audience. Based on application and performance objectives, AI interpretability demands a balance between accuracy and efficiency (Preece 2018).

TABLE 7.1
AI Interpretability Techniques

AI Interpretability Techniques	Description	Application	Strength	Weakness
Model-based techniques	Decision trees, rule-based systems, and linear models	Classification, regression	Easy to interpret	Less accurate than complex models
Post-hoc explanation methods	LIME, SHAP	Explaining complex AI models	Can explain any model	Can be computationally expensive
Hybrid approaches	Combine model-based and post-hoc methods	Balance accuracy and interpretability	More accurate than post-hoc methods, more interpretable than model-based methods	Can be complex to implement

FIGURE 7.1   AI interpretability techniques.

7.2.3 AI Interpretability Techniques

AI interpretability research and innovation have focused on understanding and explaining AI system behaviour (Stiglic et al. 2020). Decision trees, rule-based systems, and linear models are important models for studying decision-making because they provide a clear view (Kurek et al. 2023; Markus et al. 2021). Common AI Interpretability Techniques are shown in Figure 7.1. Local Interpretable Model-Agnostic Explanations (LIME) and Shapley Additive Explanations (SHAP) post-hoc explanation methods explain complex AI models' behavior after training (Madsen et al. 2022; Vieira & Digiampietri 2022; Dwivedi 2023). Model-based and post-hoc hybrid solutions, which balance accuracy and interpretability, are becoming popular (El-Sappagh et al. 2021). AI interpretability is not only a technological issue; research is also exploring its ethical and social implications. This investigation will reveal the details of each technique, their applications, strengths, and weaknesses, and address AI interpretability ethics. The objective is to decipher AI and enable its interpretable, responsible, and trustworthy use in human lives.

7.3 TECHNIQUES FOR IMPROVING AI INTERPRETABILITY

Model-based approaches are ways that are used to enhance the interpretability of AI models by including transparency into the model design itself. These techniques are used to increase the interpretability of AI models.

In most cases, these methods entail the use of models that are either simpler and/or more transparent, or the incorporation of characteristics or restrictions that make the

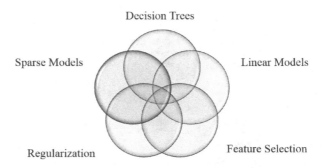

FIGURE 7.2   Popular model-based techniques.

model more amenable to interpretation. Figure 7.2 shows some popular model-based techniques.

7.3.1   MODEL-BASED TECHNIQUES

Model-based approaches have the benefit of being transparent and interpretable by design. This is because the models themselves are interpretable. On the other hand, their predictive performance may not be as good as that of more complicated models. In addition, some applications could need complicated models in order to attain the needed degree of accuracy, which makes model-based approaches less appropriate for use in these kinds of situations. Understanding transparency-enhancing AI model design strategies is essential for AI interpretability. Model-based methods use transparency into AI model creation to improve interpretability (El-Sappagh et al. 2021). These strategies use simpler, more transparent models or features and limits that make the model interpretable.

- In the world of AI models, decision trees provide a clear and understandable route from input to output. Linear models prefer simplicity and a linear relationship between input and output. The strategic lens of feature selection, also known as Feature Subset Selection, highlights the most important input qualities, simplifying the input-output relationship.
- Regularization controls model complexity by adding a penalty term to the objective function. Constraints reduce overfitting, increase generalization, and simplify the model. This technique balances interpretability and prediction performance.
- Sparse models limit their nonzero coefficients or characteristics for computational performance. Reducing the model's attributes simplifies the input-output relationship. These model-based methods improve AI model interpretability in different ways.

7.3.2   POST-HOC EXPLANATION METHODS

After an artificial intelligence model has been trained, a post-hoc explanation method is a methodology that may be used to explain the behavior of the model. When a

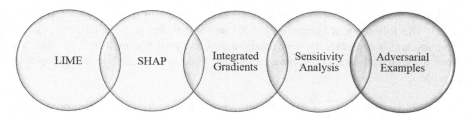

FIGURE 7.3   Common post-hoc explanation methods.

user wants to understand how the model arrived at a certain prediction or when the model is complicated and difficult to interpret, such as in the case of deep neural networks, these approaches are beneficial (Madsen et al. 2022; Vieira & Digiampietri 2022). The following are some examples of frequent post-hoc explanation methods as shown in Figure 7.3:

1. *Local Interpretable Model-Agnostic Explanations (LIME)*: LIME is an acronym that stands for "Local Interpretable Model-Agnostic Explanations." LIME creates explanations for specific predictions by fitting a simplified, interpretable model to local data points surrounding the prediction of interest. After that, the simpler model may be used to determine which characteristics contributed the most significantly to the prediction.
2. *SHapley Additive exPlanations (SHAP)*: SHAP is a uniform method for the attribution of features that may be used with any model. In addition to the capability of generating global feature significance rankings, it provides each feature with a value that indicates the extent to which it contributed to a specific prediction.
3. *Integrated Gradients*: Integrated Gradients generates the model output's gradient for each feature and integrates it from a baseline input to the actual input to determine the most important prediction factors.
4. *Sensitivity Analysis*: A sensitivity analysis identifies a model's biggest impact on projections by changing its inputs and monitoring its output.
5. *Adversarial Examples*: Adversarial examples entail purposely changing a model's input to make incorrect predictions to test its resilience. Users may understand the model's prediction process by comparing the original and adversarial inputs and examining disparities.

Post-hoc explanations may help users comprehend how AI systems forecast, but they may not fix model flaws. Specific post-hoc explanation approaches are computationally expensive, rendering them inappropriate for specific applications.

7.3.3   HYBRID APPROACHES

By integrating model-based and post-hoc explanation approaches, hybrid methods increase AI model interpretability. These approaches forecast using a complicated black-box model and employ post-hoc methods to explain its behaviour (Ali et al. 2023). The following are common hybrid techniques that may enhance prediction accuracy and interpretability over model-based and post-hoc methods, but they are more difficult and time-consuming since they need more expertise and computing expense.

1. *Interpretable Neural Networks:* Interpretable neural networks are complicated neural networks built for interpretation. They are called by their capacity to be interpreted, which may require adding layers or limitations or utilizing post-hoc approaches to explain model behavior.

2. *Rule-Based Models:* Rule-based models forecast without a sophisticated mathematical function. Experts may define these models manually or generate them using machine learning from data.

3. *Model Distillation:* Model distillation involves training a sophisticated, black-box model to forecast, then creating a smaller, interpretable model that duplicates its behavior. This may be done by training a smaller model to imitate the black-box model's outputs or utilizing post-hoc methods to explain its behavior.

4. *Surrogate Models:* Black-box models designed to replicate complex behavior are more complicated than surrogate models. They can explain or replace black-box model behavior when interpretability trumps prediction accuracy.

7.3.4 COMPARISON OF TECHNIQUES

The interpretability of AI may be improved via a variety of methods, each of which has both benefits and drawbacks. The unique application, as well as the criteria for accuracy and interpretability, should guide the selection of the approach or techniques to be used.

- The benefit of model-based procedures is that they are open to scrutiny and may be interpreted by their very nature. Typically, they include using models that are more straightforward or the incorporation of elements or restrictions that make the model more comprehensible. When compared to more complicated models, the predictive performance of these approaches may not be as good, but they are ideal for situations in which interpretability is of the utmost importance.

- The generation of explanations for the behavior of complicated black-box models is one of the tasks involved in post-hoc explanation approaches. These methods work effectively in situations in which the predicted performance of the model is the most important consideration, and in which the model has already been constructed; but they are not as clear as model-based methods.

- Hybrid methods include aspects of both model-based and post-hoc approaches, and as a result, they provide both accurate prediction and interpretability. These techniques may take more knowledge to execute and comprehend, and they may be more costly computationally than each strategy alone.

Model-based techniques, in general, are well-suited for situations in which interpretability is a priority; post-hoc explanation methods, on the other hand, are well-suited for situations in which predictive performance is a priority; and hybrid approaches, on the other hand, are well-suited for situations in which both interpretability and predictive performance are important. In the end, the decision on which technique(s) to use is determined by the unique requirements of the application as well as the balance that has to be struck between accuracy and interpretability.

7.4 LIMITATIONS AND CHALLENGES

AI interpretability faces trade-offs between interpretability and performance, scalability and complexity, human factors and cognitive biases, and regulatory and legal frameworks. Balancing interpretability and performance is crucial, and understanding the broader social, ethical, and legal implications is essential. Also, developing appropriate regulatory frameworks are crucial for safe and ethical use of AI systems. All these limitations and challenges can be discussed widely as follows:

- *Trade-Offs between Interpretability and Performance* – One of the most significant drawbacks associated with interpretability approaches is the fact that they often come at the expense of performance. When compared to more sophisticated models that are less transparent, models that are easier to comprehend could be less accurate. The trade-off between interpretability and performance is something that must be carefully examined depending on the particular application and criteria that are being met (Agarwal 2020; Jo et al. 2023; Weller et al. 2021).
- *Scalability and Complexity* – AI interpretability is complicated by scalability and complexity. Understanding how factors affect model conclusions gets harder as models become bigger and more complex, particularly deep learning models with millions of parameters (Sanz & Zhu 2021, Tambe et al. 2019).
- *Human Factors and Cognitive Biases* – Human factors and cognitive biases cause AI interpretability concerns. Even with explanations, people may fail to understand the findings, and these biases might affect how they perceive AI system explanations, leading to inaccurate conclusions (Hagendorff & Fabi 2023; Nishant et al. 2020; Choudhury 2022).
- *Regulatory and Legal Frameworks* – AI systems must be interpretable and follow data protection, security, and ethical rules. AI interpretability rules are lacking, making system installation and evaluation difficult. Increased research is required to balance performance and interpretability. Safe and ethical AI use requires examining social, ethical, and legal impacts and building regulatory frameworks (Cath 2018; O'Sullivan et al. 2019).

7.5 FUTURE DIRECTIONS

Interdisciplinary interactions along with guidelines are needed to improve interpretability as machine learning and AI evolve. The need for ethical and social implications is significant, necessitating ongoing research and regulatory frameworks. The development of explainable deep learning models and ethical guidelines is crucial for real-world applications (Choi et al. 2022; Akhai 2023).

Advances in Machine Learning and AI – The interpretability of AI will continue to be pushed forward by developments in machine learning and artificial intelligence. It will be more vital to create new methods and strategies for analyzing the behavior of AI models as these models grow more complex and smarter. In particular, there is a substantial amount of untapped potential for the creation of deeper learning models that are more explainable.

Multidisciplinary Approaches and Collaborations – The issue of interpretability is a difficult one that cuts over many different disciplines of study and need the cooperation of specialists in areas such as computer science, statistics, psychology, cognitive science, and others. It is anticipated that future study will entail more interdisciplinary approaches to interpretability, as well as more cooperation between academics, private enterprise, and public sector organizations.

Standards and Guidelines for AI Interpretability – There is a rising demand for rules and criteria around the interpretability of artificial intelligence, especially in fields such as healthcare, banking, and transportation, where the choices made by AI systems may have substantial implications in the real world. Collaborative effort from academics, policymakers, and industry stakeholders will be required in order to produce such standards and recommendations.

Ethical and Social Implications of AI Interpretability – Trustworthy, transparent, and accountable AI systems are needed due to AI's ethical and societal ramifications. This demands continual study into AI interpretability's ethical and social consequences and the creation of suitable regulatory frameworks to assure safe and ethical usage of AI systems as they gain power.

7.6 CONCLUSION

- This chapter emphasizes interpretability and performance trade-offs in AI decision-making. A sophisticated strategy that considers application needs and data complexity is needed. Modern AI models, especially deep learning systems, are scalable and complicated, making interpretability difficult. AI explanations must bridge the cognitive gap to be understandable by varied audiences.
- The chapter discusses model-based, post-hoc explanation, and hybrid interpretability methodologies. Post-hoc approaches help understand complicated models after training, whereas model-based tactics give clear interpretability routes. Understandable neural networks and model distillation are hybrid techniques that balance accuracy and interpretability. Each method has pros and cons for different applications.
- AI interpretability has ethical, legal, and societal concerns, as discussed. Safe and ethical AI deployment is hindered by unclear legal frameworks. Regulatory frameworks must change with technology to guarantee interpretability without sacrificing ethics.

- AI interpretability will benefit from multidisciplinary cooperation, norms, and ethics. Machine learning advancements, interdisciplinary alliances, and standards and norms will advance the area. Explainable deep learning models and ethical principles are needed for real-world applications, especially in healthcare and finance.

Thus, we can conclude that trustworthy and dependable AI needs continual research, cooperation, and regulatory frameworks. The path to trustworthy and dependable AI requires constant investigation and a shared commitment to create transparent, interpretable, and responsible AI systems that improve human lives and protect ethics.

REFERENCES

Agarwal, S. (2020). Trade-offs between fairness, interpretability, and privacy in machine learning (Master's thesis, University of Waterloo).

Akhai, S. (2023). *From Black Boxes to Transparent Machines: The Quest for Explainable AI. Available at SSRN 4390887.* Available at SSRN: https://ssrn.com/abstract=4390887 or http://dx.doi.org/10.2139/ssrn.4390887

Akhai, S. (2023). Healthcare record management for healthcare 4.0 via blockchain: a review of current applications, opportunities, challenges, and future potential. *Blockchain for Healthcare 4.0 (1st Edition)*, 211–223. CRC Press.

Albahri, A. S., Duhaim, A. M., Fadhel, M. A., Alnoor, A., Baqer, N. S., Alzubaidi, L., … Deveci, M. (2023). A systematic review of trustworthy and explainable artificial intelligence in healthcare: Assessment of quality, bias risk, and data fusion. Information Fusion, 96, 156–191.

Ali, S., Abuhmed, T., El-Sappagh, S., Muhammad, K., Alonso-Moral, J. M., Confalonieri, R., … Herrera, F. (2023). Explainable Artificial Intelligence (XAI): What we know and what is left to attain Trustworthy Artificial Intelligence. *Information Fusion, 99*, 101805.

Bertrand, A., Belloum, R., Eagan, J. R., & Maxwell, W. (2022). How cognitive biases affect XAI-assisted decision-making: A systematic review. In *Proceedings of the 2022 AAAI/ ACM Conference on AI, Ethics, and Society*, United Kingdom (pp. 78–91). https://hal. telecom-paris.fr/hal-03684457

Bharati, S., Mondal, M. R. H., & Podder, P. (2023). A review on explainable artificial intelligence for healthcare: Why, how, and when? *IEEE Transactions on Artificial Intelligence (Early Access)*, 1–15, https://doi.org/10.1109/TAI.2023.3266418

Carvalho, D. V., Pereira, E. M., & Cardoso, J. S. (2019). Machine learning interpretability: A survey on methods and metrics. *Electronics, 8*(8), 832.

Cath, C. (2018). Governing artificial intelligence: Ethical, legal and technical opportunities and challenges. *Philosophical Transactions of the Royal Society A: Mathematical, Physical and Engineering Sciences, 376*(2133), 20180080.

Choi, J. B., Nguyen, P. C., Sen, O., Udaykumar, H. S., & Baek, S. (2022). Artificial intelligence approaches for materials-by-design of energetic materials: State-of-the-art, challenges, and future directions. arXiv preprint *arXiv:2211.08179.* https://arxiv.org/ abs/2211.08179v2

Choudhury, A. (2022). Toward an ecologically valid conceptual framework for the use of artificial intelligence in clinical settings: Need for systems thinking, accountability, decision-making, trust, and patient safety considerations in safeguarding the technology and clinicians. *JMIR Human Factors, 9*(2), e35421.

Dwivedi, R., Dave, D., Naik, H., Singhal, S., Omer, R., Patel, P., ... Ranjan, R. (2023). Explainable AI (XAI): Core ideas, techniques, and solutions. *ACM Computing Surveys*, *55*(9), 1–33.

El-Sappagh, S., Alonso, J. M., Islam, S. R., Sultan, A. M., & Kwak, K. S. (2021). A multilayer multimodal detection and prediction model based on explainable artificial intelligence for Alzheimer's disease. *Scientific Reports*, *11*(1), 2660.

Embarak, O. (2023, May). Decoding the black box: A comprehensive review of explainable artificial intelligence. In *2023 9th International Conference on Information Technology Trends (ITT) Dubai, United Arab Emirate*, (pp. 108–113). IEEE.

Gilpin, L. H., Bau, D., Yuan, B. Z., Bajwa, A., Specter, M., & Kagal, L. (2018, October). Explaining explanations: An overview of interpretability of machine learning. In *2018 IEEE 5th International Conference on data science and advanced analytics (DSAA)* (pp. 80–89). IEEE.

Hagendorff, T., & Fabi, S. (2023). Why we need biased AI: How including cognitive biases can enhance AI systems. *Journal of Experimental & Theoretical Artificial Intelligence*, 1–14, 2203.09911.pdf (arxiv.org) [Google Scholar].

Hassija, V., Chamola, V., Mahapatra, A., Singal, A., Goel, D., Huang, K., ... & Hussain, A. (2023). Interpreting black-box models: A review on explainable artificial intelligence. *Cognitive Computation*, *16*(1), 1–30.

Jin, D., Sergeeva, E., Weng, W. H., Chauhan, G., & Szolovits, P. (2022). Explainable deep learning in healthcare: A methodological survey from an attribution view. *WIREs Mechanisms of Disease*, *14*(3), e1548.

Jo, N., Aghaei, S., Benson, J., Gomez, A., & Vayanos, P. (2023, August). Learning optimal fair decision trees: Trade-offs between interpretability, fairness, and accuracy. In *Proceedings of the 2023 AAAI/ACM Conference on AI, Ethics, and Society* (pp. 181–192).

Kurek, W., Pawlicki, M., Pawlicka, A., Kozik, R., & Choraś, M. (2023, July). Explainable artificial intelligence 101: Techniques, applications and challenges. In *International Conference on Intelligent Computing* (pp. 310–318). Singapore: Springer Nature Singapore.

Madsen, A., Reddy, S., & Chandar, S. (2022). Post-hoc interpretability for neural nlp: A survey. *ACM Computing Surveys*, *55*(8), 1–42.

Markus, A. F., Kors, J. A., & Rijnbeek, P. R. (2021). The role of explainability in creating trustworthy artificial intelligence for health care: A comprehensive survey of the terminology, design choices, and evaluation strategies. *Journal of Biomedical Informatics*, *113*, 103655.

Mehdiyev, N., & Fettke, P. (2021). Explainable artificial intelligence for process mining: A general overview and application of a novel local explanation approach for predictive process monitoring. *Interpretable Artificial Intelligence: A Perspective of Granular Computing*, *937*, 1–28.

Nishant, R., Kennedy, M., & Corbett, J. (2020). Artificial intelligence for sustainability: Challenges, opportunities, and a research agenda. *International Journal of Information Management*, *53*, 102104.

O'Sullivan, S., Nevejans, N., Allen, C., Blyth, A., Leonard, S., Pagallo, U., ... Ashrafian, H. (2019). Legal, regulatory, and ethical frameworks for development of standards in artificial intelligence (AI) and autonomous robotic surgery. *International Journal of Medical Robotics and Computer Assisted Surgery*, *15*(1), e1968.

Papadimitroulas, P., Brocki, L., Chung, N. C., Marchadour, W., Vermet, F., Gaubert, L., ... & Hatt, M. (2021). Artificial intelligence: Deep learning in oncological radiomics and challenges of interpretability and data harmonization. *Physica Medica*, *83*, 108–121.

Preece, A. (2018). Asking 'Why' in AI: Explainability of intelligent systems–perspectives and challenges. *Intelligent Systems in Accounting, Finance and Management, 25*(2), 63–72.

Reyes, M., Meier, R., Pereira, S., Silva, C. A., Dahlweid, F. M., Tengg-Kobligk, H. V., ... & Wiest, R. (2020). On the interpretability of artificial intelligence in radiology: Challenges and opportunities. *Radiology: Artificial Intelligence, 2*(3), e190043.

Saleem, R., Yuan, B., Kurugollu, F., Anjum, A., & Liu, L. (2022). Explaining deep neural networks: A survey on the global interpretation methods. Neurocomputing, (513), 165–180. https://doi.org/10.1016/j.neucom.2022.09.129

Santhoshkumar, S. P., Susithra, K., & Prasath, T. K. (2023). An overview of artificial intelligence ethics: Issues and solution for challenges in different fields. *Journal of Artificial Intelligence and Capsule Networks, 5*(1), 69–86.

Sanz, J. L., & Zhu, Y. (2021, September). Toward scalable artificial intelligence in finance. In *2021 IEEE International Conference on Services Computing (SCC)* (pp. 460–469). IEEE, Chicago, USA.

Soori, M., Arezoo, B., & Dastres, R. (2023). Artificial intelligence, machine learning and deep learning in advanced robotics, A review. *Cognitive Robotics, 3*, 54–70. https://doi.org/10.1016/j.cogr.2023.04.001

Srinivasu, P. N., Sandhya, N., Jhaveri, R. H., & Raut, R. (2022). From blackbox to explainable AI in healthcare: Existing tools and case studies. *Mobile Information Systems, 2022*, 1–20.

Stiglic, G., Kocbek, P., Fijacko, N., Zitnik, M., Verbert, K., & Cilar, L. (2020). Interpretability of machine learning-based prediction models in healthcare. *Wiley Interdisciplinary Reviews: Data Mining and Knowledge Discovery, 10*(5), e1379.

Tambe, P., Cappelli, P., & Yakubovich, V. (2019). Artificial intelligence in human resources management: Challenges and a path forward. *California Management Review, 61*(4), 15–42.

Thakur, K., Kaur, M., & Kumar, Y. (2022). Artificial intelligence techniques to predict the infectious diseases: Open challenges and research issues. In *2022 2nd International Conference on Technological Advancements in Computational Sciences (ICTACS)* (pp. 109–114). IEEE, Uzbekistan. DOI: https://doi.org/10.1109/ICTACS56 270.2022.9988282

Theis, S., Jentzsch, S., Deligiannaki, F., Berro, C., Raulf, A. P., & Bruder, C. (2023, July). Requirements for Explainability and Acceptance of Artificial Intelligence in Collaborative Work. In *International Conference on Human-Computer Interaction* (pp. 355–380). Cham: Springer Nature Switzerland.

Vieira, C. P., & Digiampietri, L. A. (2022, May). Machine learning post-hoc interpretability: A systematic mapping study. In *XVIII Brazilian Symposium on Information Systems,* May 16–19, 2022, Curitiba, Brazil (pp. 8). ACM, New York, NY, USA. https://doi.org/10.1145/3535511.3535512

Weller, D. L., Love, T. M., & Wiedmann, M. (2021). Interpretability versus accuracy: A comparison of machine learning models built using different algorithms, performance measures, and features to predict E. coli levels in agricultural water. *Frontiers in Artificial Intelligence, 4*, 628441.

8 XAI and Disease Diagnosis

Tuba Aftab, Mansoor Hussain,
M. Ahsan Saeed, Asma Yousaf,
Naseer Ali Shah, and Haroon Ahmed

8.1 INTRODUCTION

Disease diagnosis is a complex task that requires accurate and timely identification of patterns and biomarkers indicative of specific conditions. The advent of Artificial Intelligence (AI) has revolutionized healthcare, enabling the development of predictive models to aid in disease diagnosis. However, the lack of transparency and interpretability in AI models has raised concerns regarding their acceptance and reliability in clinical practice. The study of machine learning, especially computer algorithms, is known as artificial intelligence (AI) (Goldenberg, Nir, & Salcudean, 2019). Expert systems, machine vision, voice recognition, and NLP (Natural Language Processing) are a few examples of specific AI uses (de-Lima-Santos & Ceron, 2021). Predictive algorithms can now help with auto diagnosis and forecast while reducing spotting errors compared to exclusively depending on human knowledge thanks to Explainable Algorithmic Intelligence (XAI) (Krittanawong, Zhang, Wang, Aydar, & Kitai, 2017).

This chapter explores the emerging field of Explainable Artificial Intelligence (XAI) and its application in disease diagnosis. By providing transparent and interpretable explanations, XAI techniques bridge the gap between AI algorithms and human experts, improving trust, clinical decision-making, and patient outcomes. Building tools that use AI technologies to aid physicians in making wise health choices is the goal of research in medical AI (Ahmed, Mohamed, Zeeshan, & Dong, 2020). Surgery and illness detection are two examples of how AI is used in medicine. XAI is a set of approaches and tools that enables people to comprehend and value the information and output generated by Machine Learning (ML) technology (Antoniadi et al., 2021; Vilone & Longo, 2020). It helps define fairness, model correctness, transparency, and results in Machine learning decision-making. The XAI system is constructed on these three components:

8.1.1 MODEL AGNOSTIC

This approach is flexible and can be used with any kind of AI algorithm or just with a specific class of algorithms (template) (Sharma, Henderson, & Ghosh, 2020).

                    DOI: 10.1201/9781003426073-8

8.1.2 POST HOC

After the model has been taught, answers can be produced that are not based on its structural characteristics.

8.1.3 INTRINSIC UNDERSTANDABILITY

As an alternative, models can be made to be naturally simple to comprehend (Du, Liu, & Hu, 2019). Based on data trends, machine learning (ML) algorithms can forecast results. However, it can occasionally be challenging to understand or explain these forecasts. This problem is addressed by explainable artificial intelligence (XAI), which makes program results transparent and comprehensible (Arrieta et al., 2020). Contrary to conventional AI, XAI provides explanations of how a forecast was made by looking at characteristics like a unicorn's horn, mane, and tail. Humans and machines can both understand the answers provided by XAI (Páez, 2019).

8.2 XAI: AN OVERVIEW

8.2.1 DEFINITION AND CONCEPT OF XAI

Human users are able to comprehend and accept the results generated by machine learning algorithms thanks to a variety of processes and techniques collectively referred to as explainable artificial intelligence (XAI) (Dağlarli, 2020).

In recent years, artificial intelligence (AI) has advanced quickly, and it now has the ability to transform many industries (Agarwal, Swami, & Malhotra, 2022). However, for AI to have a real effect, it must be explicable, meaning that humans must be able to comprehend the logic behind its choices (Robbins, 2019). Recent developments in sub-symbolic methods, such as ensembles and deep neural networks, show promise in providing explanations, a capability lacking in earlier AI approaches (Arrieta et al., 2020). This area of study is known as explainable AI (XAI), and it is regarded as being essential for the practical use of AI models. Figure 8.1 shows the timeline of XAI disease diagnosis.

In this chapter, we examine the research on XAI, its accomplishments, and what needs to be done. We look at previous definitions of explainability in machine learning and suggest a new definition that puts the intended audience front and center. We offer a classification of recent contributions that handle the explainability of different ML models based on this concept. The intersection of data merging and explainability is one of the difficulties that XAI must overcome, and this topic has been explored in great detail in the literature. We offer a solution to these problems called "Responsible Artificial Intelligence," which places an emphasis on fairness, model explain ability, and responsibility in the extensive use of AI methods in real companies. Our goal is to not only persuade specialists and pros from other fields to use AI in their fields without any preconceived notions about its interpretability, but also to provide newbies to the field of XAI with a thorough taxonomy to aid in future research.

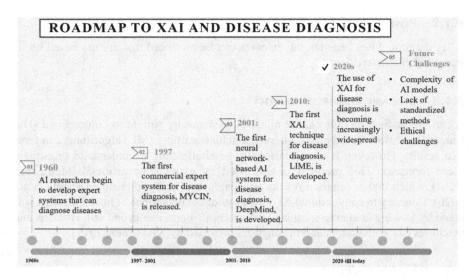

FIGURE 8.1   shows timeline depicting the evolution of disease diagnosis using Explainable AI (XAI).

8.2.2   IMPORTANCE OF XAI IN DISEASE DIAGNOSIS

Researchers in the field of medical imaging are increasingly using XAI to provide explanations for their method's results. The goal is to assist individuals in understanding the decision-making process by explaining how a neural network arrives at a specific choice. This form of explanation is highly valuable (Burrell, 2016). The use of AI has been extensive in medical applications; however, its application in clinical settings has faced challenges due to certain AI methods lacking comprehensibility. In response, researchers are actively working on developing XAI that can provide both decision-making and model explanations to address this issue.

8.2.3   ADVANTAGES OF USING OF XAI IN DISEASE DIAGNOSIS

Celiac disease (CD) is a condition in which gluten consumption leads to damage in the small intestinal mucosa among individuals with a genetic predisposition. The diagnosis of CD is challenging due to its diverse range of clinical symptoms, including atypical and asymptomatic forms. While serologic tests have been developed in the past decade to identify individuals at risk for the disease, histologic evidence of intestinal damage remains the gold standard for diagnosis. Treatment typically involves a lifelong exclusion diet that avoids gluten.

AI has enabled healthcare professionals to gain deeper insights into patient patterns and needs through extensive data analysis (Majnarić, Babič, O'Sullivan, & Holzinger, 2021). As technology advances and new medical applications are discovered, doctors and nurses will be able to offer improved guidance, support, and feedback (Lu, Xiao, Sears, & Jacko, 2005). ML in particular, and AI in general, have demonstrated significant potential across various domains (Islam, Ahmed, Barua, & Begum, 2022) examples of these include the capacity to surpass humans in difficult video games and the development of self-driving cars. ML has advanced quickly due to a number of variables, including the creation of novel statistical learning techniques, the accessibility of vast datasets, and the development of powerful hardware. By enabling software systems to learn from historical data and generate predictions based on the present situation, this advancement has increased efficiency.

8.3 TRADITIONAL DIAGNOSTIC APPROACHES

By enabling us to analyse more data and find links or insights that might otherwise go unnoticed, AI methods complement conventional methodologies and our human capacity (Hashimoto, Rosman, Rus, & Meireles, 2018). Augmented reality (AR) is a device that overlays digital data on the actual world, creating a bridge between reality and how it is perceived. The concept of *augmented reality* was first presented in 1968 as a head-mounted display, but the word wasn't created until 1992. Industries like education, building management, e-commerce, robotics, automobiles, healthcare, tourism, entertainment, and product lifecycle management could all be significantly changed by AR. By allowing direct registration of 3D real and virtual items in real time, AR can lessen brain burden. Figure 8.2 shows a flowchart representation of traditional disease diagnostics.

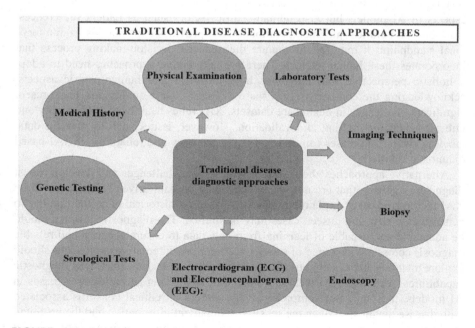

FIGURE 8.2 Medical professionals using traditional methods for disease diagnostic approaches in clinic.

8.3.1 LIMITATIONS OF CURRENT DIAGNOSTIC METHODS

- Ethical implications. There are, of course, many advantages to trusting algorithms.
- Predefined limitations
- Data paucity
- Lack of transparency
- Inconsistent outcomes

8.3.2 NEED FOR AN ALTERNATIVE APPROACH

Despite significant advancements, Explainable Artificial Intelligence (XAI) may not always offer adequate explanations for complex AI models, particularly those based on deep learning neural networks. These models tend to be intricate and challenging to interpret, leading to limitations in XAI's ability to provide transparent and trustworthy explanations. In such cases, it becomes necessary to explore alternative approaches that can address these challenges and enhance the diagnostic process. Moreover, XAI methods primarily focus on explaining predictions made by blackbox models. While these models demonstrate high accuracy, their lack of transparency poses a fundamental issue. Merely relying on XAI to interpret these models might not fully address their inherent opacity, making them less suitable for critical medical decisions. Hence, alternative approaches are essential to ensure transparency and trust in the diagnostic process.

Disease diagnosis is a multifaceted process that involves not only the technical aspects of AI models but also human-centric factors such as patient preferences, emotions, and clinical judgment. Although XAI provides explanations from a technical standpoint, it may fail to capture the nuanced decision-making process that incorporates these human factors. Therefore, alternative approaches need to adopt a holistic perspective that integrates both technical and human-centric aspects, acknowledging the complexity of disease diagnosis. Data quality and bias present significant challenges in healthcare datasets. XAI relies heavily on high-quality and unbiased data for training and validation. However, issues such as missing data, labeling errors, and biases can undermine the accuracy and reliability of XAI-based diagnostic solutions.

Alternative approaches should address these data challenges and develop robust diagnostic methods that are not solely dependent on data-driven techniques. While XAI methods excel at explaining decisions based on historical data, their ability to generalize to new and unseen cases may be limited. The diagnostic process should be adaptable and capable of learning from new data to ensure accurate and reliable diagnosis across a wide range of scenarios. Therefore, alternative approaches should explore methods that can effectively adapt and utilize new data, enhancing diagnostic capabilities. Although XAI techniques assist in identifying and mitigating biases in AI models, they may not comprehensively address all ethical concerns associated with disease diagnosis. Given the sensitivity of patient information and the profound implications for their lives, alternative approaches should actively consider and address ethical considerations. These include privacy, security, informed consent, and fair distribution of healthcare resources, ensuring that the diagnostic process upholds the highest ethical standards.

In conclusion, while XAI has made significant strides in disease diagnosis, alternative approaches are necessary to overcome limitations related to complex AI models, black-box opacity, human-centric factors, data challenges, generalization, and ethical considerations. By combining technical expertise with a comprehensive understanding of the human element, these alternative approaches can foster transparency, trust, and accuracy in the diagnostic process, ultimately leading to improved patient outcomes.

8.4 XAI TECHNIQUES FOR DISEASE DIAGNOSIS

8.4.1 MACHINE LEARNING

Machine learning (ML) is now widely used in modern technologies like robotics, computers, and smartphones as well as in healthcare for illness detection and safety (Chattu, 2021). Machine-learning-based disease diagnosis (MLBDD), which provides time and expense savings, is one field of special promise (Ahsan, Luna, & Siddique, 2022). Traditional diagnostic procedures are frequently costly, time-consuming, and constrained by human capability; in contrast, MLBDD systems are unrestricted and are capable of handling massive amounts of patient data (Purohit & Mustafa, 2015). They become particularly helpful in healthcare environments with a high patient burden because of this. Healthcare data, including tabular data (such as patients' medical records, age, and gender) and medical pictures (such as X-rays and MRIs), are used to build MLBDD systems.

Machine learning has significantly improved healthcare by allowing medical workers to quickly and correctly diagnose illnesses (Ghazal et al., 2021). It entails the study of medical data, such as patient histories, physical exam findings, laboratory test results, and images obtained through medical imaging, in order to spot patterns that point to the existence or lack of a specific illness. Large volumes of data can be rapidly and precisely analyzed by ML algorithms, which is advantageous in areas like radiology. Algorithms educated on large databases of medical images can identify patterns and abnormalities that may be challenging for humans to discover. Physicians must evaluate immense sets of medical images to find anomalies or indications of illness. In order to achieve improved therapy results, early illness diagnosis is crucial.

ML algorithms can assist by finding patterns in vast amounts of medical data. The algorithms can identify trends that signify a specific disease's early phases, allowing doctors to take preventive measures and start treating patients right away (Groves, Kayyali, Knott, & Kuiken, 2016). To give a thorough picture of a patient's health, ML can also assist doctors in making more accurate assessments by evaluating multiple sources of medical data, such as, patient histories, laboratory tests, and medical images. As a result, assessments are more accurate and treatments are more efficient.

8.4.2 DEEP LEARNING

In the realm of medical diagnosis, computer-aided diagnosis and detection is a quickly expanding study topic (Zameer, Tariq, Noreen, Sadaqat, & Azeem, 2022). It is especially helpful for people who reside far from medical facilities, cannot pay for medical expenses, or cannot take time off from work (Bodenheimer, 2008). For the detection and diagnosis of illnesses like Alzheimer's disease, cancer, rheumatoid arthritis, liver disease, diabetic retinopathy, lung disease, heart disease, hepatitis, dengue, and Parkinson's disease, computer scientists have created numerous AI programs. ML-based disease detection accuracy improvements have been the subject of a recent study. Deep learning systems, in particular, have shown tremendous potential in illness detection and diagnosis (Mishra, Dash, & Jena, 2021).

In this chapter, we go over current trends and advancements in deep learning and how they might affect the early detection and diagnosis of different illnesses. The area of medical imaging has seen special success with deep learning. Deep learning models can reliably identify anomalies and make diagnoses of illnesses by being trained on huge databases of medical pictures (De Bruijne, 2016). For instance, using medical pictures, deep learning techniques have been utilized to distinguish breast cancer, lung cancer, and diabetic retinopathy (Golden, 2017). Early identification and treatment of prospective problems are made possible by the models' fast and precise analysis of medical images.

Analysis of genomic data is another field where deep learning can be helpful for illness diagnosis. These algorithms can find genetic mutations that might be indicative of particular illnesses by training them on big genomic data sets. For instance, BRCA1 and BRCA2 gene mutations, which are linked to a higher chance of breast and ovarian cancer, have been identified using deep learning algorithms (Dees et al., 2012).

8.4.3 DECISION TREES

One alternative approach that can enhance the transparency and interpretability of ML models in disease diagnosis is the use of decision. Decision trees provide a structured representation of the decision-making process employed by the ML program, making it easier for medical personnel to recognize the most crucial characteristics used in the diagnosis and comprehend how the model arrives at its findings. In the following we discuss how decision trees can contribute to the understanding of disease diagnosis.

8.4.3.1 Clear Decision Path

Decision trees present a clear and intuitive decision path by recursively partitioning the input data based on relevant features. Each split in the tree represents a decision point based on a specific feature, allowing medical personnel to trace the decision-making process step by step. This transparency helps in understanding the key factors influencing the diagnosis and builds trust in the model's decisions.

8.4.3.2 Feature Importance

Decision trees explicitly display the importance of different features in the diagnosis process. The structure of the tree reflects the hierarchy of feature importance, with the most influential features appearing at the top. Medical personnel can identify and focus on these critical features, gaining insights into the diagnostic indicators that carry the most weight in the model's predictions.

8.4.3.3 Interpretability

Decision trees offer a highly interpretable representation of the decision-making process. The rules and conditions at each node of the tree can be easily understood and communicated to medical professionals and patients. This interpretability facilitates discussions among healthcare teams, enabling collaborative decision-making and improving overall diagnostic accuracy.

8.4.3.4 Handling Complex Interactions

Decision trees can handle complex interactions among different features, capturing nonlinear relationships that may exist in the data. By splitting the data based on multiple features simultaneously, decision trees can uncover intricate patterns and dependencies that contribute to accurate disease diagnosis. This capability is particularly useful when dealing with complex diseases where multiple factors interact.

8.4.3.5 Visual Representation

Decision trees can be visually represented, making it easier to communicate and comprehend the decision-making process. The tree's structure, branches, and leaf nodes can be visually displayed, providing a comprehensive overview of the diagnostic process. This visual representation aids medical personnel in understanding and validating the model's decisions.

8.4.3.6 Rule Extraction

Decision trees can also be used to extract rules or decision pathways that can be translated into actionable guidelines or clinical decision support systems. *Figure 8.3 shows flowchart which will help in understanding of decision trees in a disease diagnosis.* These rules can be shared with medical professionals, helping them make informed decisions based on the model's recommendations and promoting consistent and standardized diagnostic practices. In summary, decision trees offer a transparent and interpretable approach to disease diagnosis. They provide a clear decision path, highlight feature importance, promote interpretability, handle complex interactions, offer visual representations, and enable rule extraction. By incorporating decision trees into the diagnostic process, medical personnel can better understand and trust the ML models, leading to improved patient care and outcomes.

8.4.4 LIME (Local Interpretable Model-Agnostic Explanations)

ML uses the LIME (Local Interpretable Model-Agnostic Explanations) method to give justifications for model forecasts (Ribeiro, Singh, & Guestrin, 2016). LIME can

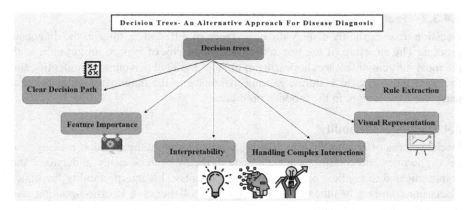

FIGURE 8.3 Decision trees: an alternative, effective approach for disease diagnosis understanding.

offer information about a patient's potential susceptibility to a specific illness or health problem in the setting of disease diagnosis. For instance, LIME can help shed light on which aspects of a mammography or other type of medical picture are most crucial in determining whether a patient has breast cancer (Williams & Jeanetta, 2016). As a result, assessments and suggestions for therapy might be more precise as doctors and other healthcare professionals comprehend the underlying factors that contribute to the illness' development. Another illustration is the use of LIME in the detection of diabetic retinopathy, a disease that can cause diabetics to go blind. LIME can be used to determine which retinal image characteristics are most crucial for determining the seriousness of the disease. The knowledge gained from this information can assist medical professionals in choosing the best course of action for their patients and in comprehending the fundamental processes that underlie illness development.

8.4.5 Shapley Additive Explanations

A ML technique called Shapley Additive Explanations (SHAP) can aid in disease detection by explaining the forecasts produced by a model (Wang et al., 2021). Each feature in the incoming data is given a number by SHAP, showing how much that feature added to the forecast. This enables medical professionals to recognize the characteristics that are most crucial for a prognosis and can shed light on the disease's fundamental causes. SHAP can be used, for instance, to diagnose Alzheimer's illness. To determine whether a new patient has Alzheimer's disease, a deep learning algorithm can be taught on a dataset of patient brain images and related diagnoses. Clinicians can use SHAP to look over the model's forecasts and determine which aspects of the brain scan are most crucial for identification. This can assist in the early identification and therapy of Alzheimer's disease by enabling doctors to recognize patterns of brain atrophy that are characteristic of the condition. In order to assist in illness detection, SHAP can also be used to analyze electronic health records (EHRs).

For example, SHAP can be used to examine EHR data and determine which patient traits and aspects of medical history are most crucial in determining the probability of getting diabetes. By implementing preventive measures, doctors can identify patients who are highly susceptible to the illness and lower their risk.

8.4.6 Artificial Neural Network

By examining vast quantities of medical data and finding patterns that might be suggestive of specific diseases, artificial neural networks (ANNs) can aid in disease detection. To spot possible problems, help with early identification, and enhance patient outcomes, ANNs are taught on sizable databases of medical images, electronic health records (EHRs), and genomic data. Here are some instances of how three distinct diseases can be diagnosed with the aid of ANNs.

8.4.6.1 Alzheimer's Disease

ANNs can be taught on MRI scans of the brain to recognize Alzheimer's disease symptoms. These models can identify structural alterations in the brain, such as shrinkage, which may be symptoms of the illness. Additionally, ANNs can be taught on EHRs to recognize signs of cognitive decline such as changes in language and memory that could be symptoms of Alzheimer's disease.

8.4.6.2 Skin Cancer

ANNs can be taught on pictures of skin lesions to recognize melanoma and other skin cancer symptoms. These models can assess the probability of a skin lesion being cancerous by examining characteristics like size, shape, and color. Additionally, ANNs can be taught on EHRs to recognize skin cancer risk factors like sun exposure and family history.

8.4.6.3 COVID-19

ANNs can be taught to recognize COVID-19 symptoms like ground-glass opacities and consolidation on chest X-rays and CT scans. These models can assist in the early diagnosis and detection of the illness, which can help with efficient treatment and stop the virus's spread. In order to forecast patient outcomes and recognize risk factors for severe COVID-19 infections, such as age and underlying medical conditions, ANNs can also be trained on EHR data.

8.4.7 Support Vector Machine

Support vector machines (SVMs) have gained popularity in the field of ML, serving as effective tools for classification, regression, and novelty detection. SVMs create a hyperplane in a high- or infinite-dimensional space to separate data into different categories and perform classification tasks. They are closely related to classical multi-layer perceptron neural networks.

The main objective of SVMs is to strike a balance between minimizing training set errors and maximizing the margin, resulting in optimal generalization while avoiding overfitting. Developed by Vapnik, SVMs have found applications in various domains, including pattern recognition, bioinformatics, and cancer diagnosis.

SVMs function by mapping data into a higher-dimensional feature space, allowing for the categorization of data points even when they are not linearly separable. By employing nonlinear mapping techniques, SVMs identify an optimal linear hyperplane that maximizes the separation margin between two classes.

In the medical field, SVMs have been extensively utilized for cancer diagnosis, including breast cancer, lung cancer, and prostate cancer. They can be trained on medical imaging data such as mammograms and CT scans to detect cancerous tissue. Additionally, SVMs have been employed to predict patient outcomes and identify risk factors for various diseases based on electronic health record (EHR) data. For example, SVMs have been employed to assess the risk of hospital readmission in patients with heart failure using medical history and clinical data. Moreover, SVMs have proven valuable in analyzing genomic data to identify genetic mutations associated with diseases like ovarian cancer.

8.5   XAI APPLICATIONS IN DISEASE DIAGNOSIS

8.5.1   CANCER DIAGNOSIS

Explainable artificial intelligence (XAI) has a significant potential to revolutionize the field of cancer diagnosis by providing interpretable and transparent models that can assist clinicians in making more accurate and reliable diagnoses. Here are some points of XAI applications in cancer diagnosis.

8.5.1.1   Automated Image Analysis

Using XAI, machine learning algorithms can be created that evaluate medical pictures like MRIs, CT scans, and mammograms to find and categorize tumors (Van der Velden, Kuijf, Gilhuijs, & Viergever, 2022). By highlighting the areas of the picture that were crucial in making the diagnostic, these models can describe how they arrived at their forecasts. Utilizing ML techniques, automated image analysis examines medical pictures in order to find tumors or other anomalies. However, it can frequently be challenging to understand the forecasts made by these algorithms, which can cause uncertainty and possibly incorrect diagnoses. XAI steps in at this point. ML models can be created using XAI that not only make precise forecasts but also explain how those predictions were made.

By highlighting the areas of the picture that were crucial in making the diagnosis, XAI can assist doctors in understanding why a specific tumor was labeled as cancerous or non-cancerous in the context of cancer detection. This may result in more precise and trustworthy diagnoses and eventually improved patient results. Additionally, XAI can assist in addressing a few issues with automated image analysis's use in cancer detection, such as the potential for errors in the data used to train ML models. Clinicians can use XAI to spot any biases or errors in the models' forecasts and correct them by giving explanations for them.

8.5.1.2 Risk Assessment

XAI models have the potential to assist healthcare professionals in assessing a patient's risk of developing cancer (Payrovnaziri et al., 2020). This can be done by analyzing various patient data such as medical history, genetic information, and lifestyle factors. By doing so, the models can identify the features that contribute the most to the risk assessment and provide an explanation for their predictions. For example, a ML model trained to predict breast cancer risk might identify factors such as age, family history, and genetic mutations as the most important features. One of the advantages of XAI models is their ability to explain how they arrived at their predictions. By identifying the factors that were most influential in the decision-making process, clinicians can better understand the rationale behind the predictions and identify any biases or inaccuracies in the model. This can help in prioritizing interventions and treatments that address the most significant risk factors. Moreover, XAI models can be used to develop personalized risk assessments that consider a patient's unique medical history, genetic profile, and lifestyle factors. This approach can lead to the development of more effective prevention and treatment strategies for cancer.

8.5.2 TREATMENT PLANNING

Here are some ways in which XAI can be helpful in this context:

8.5.2.1 Analysis of Medical Records

Large datasets of patient medical records, which contain details like the sort of cancer, its stage, and previous treatments, can be used to create XAI models. These data can be examined by XAI models to find trends and connections between patient characteristics and therapeutic results.

8.5.2.2 Prediction of Treatment Response

The likelihood of a patient responding a certain way to various therapy choices can be predicted using XAI models. Based on a patient's genetic make-up and medical history, a model might, for instance, forecast that the patient will react better to chemotherapy than radiation treatment.

8.5.2.3 Optimization of Treatment Plans

By determining the most effective set of therapies for a given patient, XAI models can aid in the optimization of treatment plans. XAI models can assist doctors in creating more individualized and successful treatment plans by taking into account elements like cancer type, stage, and patient traits.

8.5.2.4 Explanation of Treatment Recommendations

By emphasizing the variables that affected the choice, XAI models can explain why a specific course of therapy was advised. Patients and medical professionals who use this information can make better choices about their care by better understanding the reasoning behind the treatment plan. Overall, XAI has the potential to be an effective instrument for creating tailored therapy strategies for cancer patients. XAI models can assist doctors in delivering more effective and personalized care for cancer patients by analyzing medical records, forecasting treatment reactions, improving treatment plans, and outlining treatment suggestions (Roy, Meena, & Lim, 2022).

8.5.3 CLINICAL DECISION SUPPORT

Here are some ways in which XAI can be helpful in this context.

8.5.3.1 Analysis of Patient Data

Large datasets of patient data, including details like a patient's medical background, genetic information, and therapy results, can be used to teach decision support systems. These data can be analyzed by XAI models to find trends and correlations that can guide therapy choices.

8.5.3.2 Prediction of Outcomes

To forecast how a patient will likely react to various therapies, XAI models can be used. For instance, a model might forecast that a patient with a specific cancer type and DNA mutation will likely react better to a particular kind of treatment.

8.5.3.3 Recommendation of Treatment Options

XAI models can offer suggestions for therapy choices based on the analysis of patient data and forecasts of treatment results. These suggestions may take into account the traits of the patient, the sort and stage of the disease, and genetic information.

8.5.3.4 Explanation of Recommendations

By emphasizing the variables that affected the choice, XAI models can explain why a specific course of therapy was advised. This can assist clinicians in comprehending the reasoning behind the suggestion and informing their choices regarding the treatment of patients.

8.5.4 INFECTIOUS DISEASE DIAGNOSIS

8.5.4.1 Image Analysis and Artificial Intelligence in Infectious Disease Diagnostics

Because they have the necessary image analysis skills, such as pathogen recognition, inflammatory context evaluation, and colony development classification on agar plates, microbiologists play a vital part in clinical microbiology laboratory diagnostics. But new advances in AI have made it possible to automate these processes, which could greatly improve the precision and promptness of evaluations. Gram stains, fecal and blood smears, and histopathologic slides, which are crucial in identifying the presence and type of microorganisms, host inflammatory reaction, and specimen quality, could all be successfully interpreted by AI-powered picture analysis. Additionally, automating the categorization of colony development on agar plates, which necessitates the discovery of possible pathogens and normal flora, could greatly enhance microbial evaluation. Automating these visually interpretive duties could be a game-changer for the clinical microbiology laboratory due to a persistent lack of medical laboratory scientists. Fortunately, the development of the most recent AI algorithms that can accurately distinguish images has made it possible to automate clinical microbiology analysis, which could improve productivity and diagnostic precision (Visibelli, Roncaglia, Spiga, & Santucci, 2023).

8.5.4.2 Artificial Intelligence as a Fundamental Tool in Management of Infectious Diseases

Pathogens like bacteria, viruses, fungi, or parasites can cause infectious illnesses, which can either be symptomatic or silent. The negative effects of infectious illnesses, however, are not entirely explained by symptoms alone because some, like HIV, can have serious negative effects even if they start out as silent. AI programs have had a significant effect on screening, analysis, forecast, and monitoring of sick people and have emerged as crucial tools for spotting early indicators of contagious illnesses. AI can aid with disease management, medical support, radiology imaging, computational tomography, database creation, and the generation of data on verified, recovered, and mortality cases. Global pandemics like SARS, MERS, Ebola, and Zika have had a significant impact on humanity in the twenty-first century. Convolutional neural networks, quick diagnosis methods, and other AI techniques are being used to handle COVID-19 and distinguish SARS-CoV-2 positive and negative groups. These initiatives cover the creation of vaccines, illness management, diagnosis, patient records, prevention of spread, societal exclusion, and forecasting upcoming pandemics (Jaycox et al., 2023).

Despite advances in medicine, infectious illnesses like the Spanish Flu, T, B, Ebola, SARS, and influenza have been among the leading causes of mortality worldwide, particularly in low-income countries. The recent global outbreaks of Ebola, SARS, and influenza demonstrate how dangerous these infectious illnesses have grown to be for people. Researchers and healthcare partners have devised innovative methods to support the healthcare industry, particularly in tracking the spread of infectious diseases. Machine learning and AI have revolutionized disease management by simulating virus behavior and developing novel containment strategies. Predictive modeling, utilized by health experts, enhances existing interventions to halt the transmission of contagious diseases. Governments worldwide collaborate with local authorities and healthcare workers to monitor, respond to, and mitigate the spread of infectious diseases.

Advancements in mathematical tools enable scientists to accurately predict disease spread, considering the unique characteristics of each pathogen and establishing vaccination distribution goals. Leveraging vast amounts of data, including health information, human behavior patterns, and environmental factors, big data analytics prove valuable in responding to deadly outbreaks. The COVID-19 pandemic has resulted in a number of innovations, including over-the-counter infectious illness testing, mass spectrometry-based COVID-19 host response detection, and the use of AI and ML to diagnose SARS-CoV-2 infection. These groundbreaking methods will continue to have a big influence even after the present epidemic is over, ushering in a new age of infectious disease testing that goes beyond SARS-CoV-2 diagnosis.

8.5.4.3   Point-of-care Testing

Point-of-care (POC) testing refers to medical exams done on or near the patient. This type of investigation has historically been conducted using benchtop, handheld, portable, and transportable tools. Significant improvements in POC testing, especially in the area of infectious disease detection, have been made recently. This includes the development of rapid molecular testing that may be performed at the patient's bedside and the growing use of mobile health tracking devices. Over-the-counter (OTC) and direct-to-consumer (DTC) trials have also expanded.

8.5.4.4   Rapid Molecular Testing

Molecular infectious disease testing has undergone a transformation due to the speed and mobility improvements brought about by automation and miniaturization. Tests that used to take days to complete are now finished in a matter of hours or even minutes. Rapid molecular testing is now frequently used as a consequence, particularly in emergency rooms where fast findings are required. With numerous point-of-care (POC) reverse transcription real-time PCR and isothermal nucleic amplification techniques getting emergency use authorizations (EUA) from the US Food and Drug Administration, the COVID-19 pandemic has further emphasized the significance of fast testing (FDA). New testing options for COVID-19 have also been made possible by CRISPR gene editing technology, which is also similar to RT-PCR in terms of expense and speed of SARS-CoV-2 identification without amplification. A CRISPR-based SARS-CoV-2 assay was linked to a smartphone in a research offering a limit of detection of about 100 viral copies/L with a response time of less than 30 minutes. Other uses for CRISPR-based infectious testing could include merging with flow-based tests to offer low-cost, high-throughput testing at the point of service.

8.5.5   Antimicrobial Resistance Testing

Around 2014, mass spectrometry (MS) made its way into the field of clinical microbiology(Hou, Chiang-Ni, & Teng, 2019). Many universities are now using matrix-assisted laser desorption ionization (MALDI) – time of flight (TOF) – MS to speed up the identification of bacteria and fungi directly from microbiological culture. A proteomic spectrum depicting the ionizable proteins unique to different bacterial and fungi species is produced by these MALDI-TOF-MS methods (Haider, Ringer, Kotroczó, Mohácsi-Farkas, & Kocsis, 2023). Recent advancements in this field have included the quick detection of antibiotic resistance using MALDI-TOF-MS. MALDI-TOF-MS has detected beta-lactamase activity using methods designed to assess ertapenem resistance in Bactericides fragile isolates.

8.5.6   Machine Learning Applications for Infectious Disease Testing

The area of computer science known as AI works to create tools that can mimic human behavior. A branch of AI called ML creates systems that work better after being taught on fresh data. Once considered science fiction, AI and ML have changed the world and are predicted to continue doing so quickly and possibly in unexpected ways. Examples of contemporary AI/ML applications include helping people conduct internet searches and helping businesses anticipate consumer requirements. Autonomous cars can also mimic human driving behavior. In a similar vein, and as has already been demonstrated in this piece, there are numerous uses that have the potential to upend the field of infectious disease testing.

8.5.6.1 Lyme Disease

Borrelia burgdorferi is known to cause Lyme disease. Early detection and diagnosis are crucial in preventing disease progression (Loomba et al., 2023). However, identifying the disease in the early stages is difficult because of the subtle signs that are often overlooked by both patients and healthcare professionals. The current Lyme disease testing methods exhibit poor sensitivity, which leads to underdiagnoses and overdiagnoses in some cases. Artificial intelligence/ML (AI/ML) was proposed to improve the performance of a point-of-care (POC) sero-diagnostic test that targeted bacterial antigens, including OspC, BmpA, P41, ErpD, Crasp1, OspA, DbpB, VlsE, P35, and Mod-C6. The ML algorithm was able to achieve a sensitivity of 90.5% and specificity of 87.0% with conventional serology, which is significantly higher than the current testing methods. By combining POC serology with AI/ML, it is possible to improve the accuracy of diagnosis and provide timely treatment, leading to better outcomes for patients.

8.5.6.2 Meningitis

With about 36,000 hospitalizations per year recorded in the United States, meningitis continues to have a major financial impact on healthcare (Versfeld, 2023). Although molecular tests have increased the speed of finding meningitis-causing pathogens, cerebrospinal fluid continues to be the major specimen type for diagnosis. In addition to detecting the amounts of white blood cells, glucose, and protein, the Gram Stain test for meningitis also needs a sample of the patient's CSF. AI and ML techniques have been employed to predict meningitis by utilizing non-cerebrospinal fluid (CSF) parameters, circumventing certain limitations. A neural network-based model utilized six characteristics, including lymphocyte count, blood glucose, and age, achieving a testing accuracy of 86.3%. ML methods were used to develop a model that exhibited a sensitivity of 99% and specificity of 100% when compared to conventional microbiological methods. This model utilized features such as age, ethnicity, sex, white blood cell count, blood glucose level, and CSF parameters like glucose, protein, and leukocytes.

8.5.6.3 Sepsis

Severe medical conditions like sepsis, which can be fatal, need to be treated right away (Patel et al., 2023). Sepsis must be detected early in order to be treated effectively, and AI/ML can be very useful in identifying people who are at danger. A plethora of patient information is available from electronic medical records (EMRs), and this information can be used to build AI/ML models that can spot patients who might be septic. Multiple ML studies have been conducted in the intensive care unit (ICU) population, utilizing various parameters such as age, gender, blood pressure, heart rate, temperature, oxygen saturation, respiratory rate, white blood cell count, microbiological culture results, lactate, high sensitivity C-reactive protein, procalcitonin, arterial blood gas measurements, use of vasopressors, and use of antibiotics as features. These studies aimed to achieve early detection of sepsis, with one study demonstrating a sensitivity/specificity of 87%. However, certain sepsis subgroups, such as burn victims, may exhibit different response patterns, challenging the effectiveness of such models. To address this issue, specifically focused on predicting the occurrence of sepsis in individuals with burns using AI/ML models. This specialized model achieved a sensitivity of 95.8% and specificity of 87.8%. The variables incorporated in this model included heart rate, body temperature, hemoglobin levels, blood urea nitrogen, and total CO2. This exemplifies the ability of AI/ML to develop tailored models for specific patient groups when new data becomes available.

8.5.7 MOLECULAR HOST-RESPONSE ANALYSIS

Molecular host-response analysis for infectious illness diagnostics can use ML techniques (Wang et al., 2023). This method includes examining how host genes or proteins are expressed in reaction to an infection, which can reveal important details about the kind and severity of the infection. Researchers can create predictive models that can precisely detect the presence of particular pathogens or distinguish between various infections by analyzing these molecular markers using ML techniques. These models can increase the efficacy and efficiency of infectious disease testing, which could eventually result in more precise and efficient therapies. Molecular host-response analysis is one way ML is used in viral disease diagnostics. Analyzing the expression of particular genes in reaction to an illness enables researchers to pinpoint the precise organism that caused it. The intricate data produced by these tests can be analyzed using ML algorithms to find trends that can be used to make precise predictions. Machine learning is used to identify individuals with bacterial or viral infections using a 29-host-mRNA 30-minute point-of-care test (Tran, Albahra, Rashidi, & May, 2022). The test's precision in finding the precise pathogen accountable for the infection was demonstrated by its area under the ROC curve of 0.92, which was accomplished using a neural network ML method. By offering quick and precise diagnoses, this kind of technology has the potential to change the testing for contagious illnesses. This could help to better patient outcomes and stop the spread of infectious diseases (Jain et al., 2021).

8.5.8 ALTERNATIVE SAMPLE TYPES

The integration of alternative sample categories in ML applications for infectious disease testing can significantly increase the availability of samples for training and validation. By incorporating different sample types, ML models can benefit from a larger dataset, resulting in improved sensitivity and precision. Traditionally, the collection of fluids and tissue specimens has been the primary approach for infectious disease testing (Lagathu et al., 2023). In the case of common respiratory infections, the nasopharyngeal (NP) swab sample has been widely accepted as the standard for testing. However, the emergence of the COVID-19 pandemic has necessitated the exploration of alternative sample types due to swab shortages and challenges associated with obtaining NP samples.

8.5.9 SALIVA SPECIMENS

Currently, several institutions have adopted the use of saliva samples for bulk screening of COVID-19 (Nagura-Ikeda et al., 2020). Rutgers University was the first organization to receive Emergency Use Authorization (EUA) for a saliva-based collection kit. Recent research and the guidance provided by the Infectious Disease Society of America (IDSA) suggest that the efficacy of saliva samples may be comparable to that of nasopharyngeal (NP) swab samples. However, the widespread use of saliva samples is somewhat limited due to the lack of a standardized diagnostic method and the incompatibility of certain EUA-approved SARS-CoV-2 PCR tests with this sample type. Pre-analytical variables in saliva specimens, such as patient food consumption prior to testing, drug use, and hydration state, among other confounding variables, may also cause inaccurate results (Anvari et al., 2021). Saliva samples have been subjected to ML methods for the diagnosis of viral illnesses like COVID-19 and the flu. In comparison to conventional sample types like nasal swabs and nasopharyngeal aspirates, saliva has a number of benefits, including simplicity of collection and patient satisfaction. In one study, a ML algorithm that was trained on data from more than 1,000 patients was able to identify COVID-19 from saliva samples with a sensitivity of 94% and a specificity of 98% (Rosado et al., 2021). This demonstrates the possibility of different sample kinds of machine learning to enhance the detection and treatment of infectious diseases.

8.5.10 BREATH SPECIMENS

Machine learning methods have also been used to diagnose tuberculosis (TB) from breath samples. A sensor array based on nanomaterials and ML methods was used to create a breath-based TB screening method [42]. The sensor array was able to identify volatile organic molecules (VOCs) unique to TB illness in breath samples. The sensitivity and precision of the ML algorithm's classification of TB-infected individuals were 93.3% and 96.4%, respectively. This technique provides a non-invasive, quick, and affordable way to test for TB, especially in settings with restricted resources.

8.5.11 CHALLENGES AND BARRIERS

The improvements in infectious disease testing addressed in this chapter should be approached cautiously because the data from earlier studies or under controlled circumstances might not accurately represent performance in the real world. For instance, MS's antimicrobial susceptibility testing is a promising technique, but it is still unknown how the intricacy of the bacterial proteome affects in vitro antimicrobial resistance in clinical settings. Additionally, the FDA's EUA route sped up the implementation of several tests during the COVID-19 pandemic. However, some tests' results in the real world did not match that outlined in the EUA that was submitted, showing less sensitivity and specificity. The data used to teach the algorithms in the area of ML is essential to their effectiveness The ML-enhanced COVID-19 MALDI-TOF-MS method mentioned in the reference paper had been evaluated using a relatively small sample size (Tsai et al., 2022). In order to undergo regulatory review, it requires completion of large-scale multicenter studies. Despite showing promising performance, further research and validation on a larger scale are necessary. The importance of secondary and even tertiary databases is emphasized by the researchers. Multicenter studies are currently being conducted for these new COVID-19 sample categories, including saliva and breath. Verifying the matrix impacts of these sample classes will take more effort.

8.6 XAI CARDIOVASCULAR DISEASE DIAGNOSIS

Over the past ten years, uses of artificial intelligence in cardiovascular studies have grown in popularity. AI algorithms have been widely employed for phenogrouping, patient prognostication, picture segmentation and reconstruction, image quality management, and gaining scientific understanding (Benjamins et al., 2022). Demographic and co-morbidity information about the patient has been used to enhance the effectiveness of ML systems. In the area of cardiology, AI-based software gadgets and risk evaluation tools have also been adopted (Albahri et al., 2023).

8.6.1 ELECTROCARDIOGRAPHY

A frequently used diagnostic instrument for determining cardiovascular problems is the electrocardiogram (ECG), but its analysis can be challenging and time-consuming (Hussain, Tariq, & Gill, 2023). AI techniques are being used more and more to help with patient stratification and prognostication in order to improve the precision of automatic ECG reading. A number of ECG characteristics, including heart rate, cardiac axis, and irregular rhythms like atrial fibrillation (AF), can be recognized by ML algorithms. Deep neural networks (DNNs) and support vector machines (SVMs) are effective at correctly identifying ECG markers of cardiac patients, including arrhythmias, according to recent research.

It is crucial to diagnose subclinical AF because it is a prevalent arrhythmia that can result in seizures and early mortality. In a historical analysis of more than 600,000 patients, a risk prediction model using neural networks was able to distinguish between patients with AF and those without it with an area under the curve (AUC) of 0.87. Convolutional neural networks (CNNs) were used in a research to check ECGs for asymptomatic paroxysmal atrial fibrillation (AF) in people with regular heartbeats (Prabhakararao & Dandapat, 2022). Over 450,000 ECGs from over 120,000 people made up the study's big collection. SVM, gradient boosting machine (GBM), and multilayer neural networks (MLNN) were used in a retrospective cohort analysis of non-valvular AF patients to forecast the risks of stroke, major bleeding, and mortality in comparison to clinical risk ratings determined by AUC. When compared to CHA2DS2-VASc, GBM demonstrated a slight performance increase for stroke, a substantial performance improvement when detecting major bleeding, and better mortality prediction (Lu et al., 2022).

Current diagnostic techniques can make it challenging to identify the disease known as atrial AF. Smartphones, smart bands or watches, earlobe monitors, and portable electrocardiogram (ECG) devices are just a few examples of the mobile devices that have been developed recently to identify AF (Prieto-Avalos et al., 2022). These tools are simple to use, allow continuous monitoring and personalized ECG signal analysis, and provide secure, non-invasive, instant access to patients.

Findings from the Apple Heart Study regarding the detection of people with asymptomatic paroxysmal AF were also encouraging. Smartphones were used in the research to identify 0.5% of patients with potentially irregular pulses, of whom 34% had an ECG that revealed they had AF. The group that received notice of their irregular pulse had a greater likelihood of starting anticoagulant or antiplatelet therapy. Additionally, a sizable portion of patients who were identified received different therapies like cardioversion, implantable loop recorders, anti-arrhythmic drugs, and ablation.

8.6.2 Transthoracic Echocardiography

For real-time monitoring of the heart and the detection of any structural flaws, echocardiography is a frequently used imaging method (Little et al., 2023). AI can improve imaging assessment precision by reducing inter- and intra-operator variability and providing additional information that might not be visible to the human eye (Sarno et al., 2022). Transthoracic echocardiography has made significant use of ML methods for image segmentation, image diagnosis, and patient prognostication (Xu, Yu, Zhang, & Zhang, 2022).

The left ventricular ejection fraction (EF), a gauge of contractile function, was autonomously determined by one cutting-edge two-dimensional echocardiographic image analysis device using AI-learned pattern recognition. This technique had less variability than ocular EF and gave findings that were comparable to the usual manual calculation. A completely automatic computer vision program (AutoLV) with ML-enabled picture analysis was able to quantify average biplane longitudinal strain (LS), left ventricular volumes, and EF in a multicenter trial with a 98% success rate and an average analysis time of 8 seconds per patient (Della Bella et al., 2022).

Using clinical and ultrasound data, ML algorithms have also been used to differentiate between constrictive pericarditis and restrictive cardiomyopathy (Dell'Angela & Nicolosi, 2022). With an AUC of 89.2%, the associative memory classifier (AMC) was discovered to be the most effective method. This technique distinguished between these two comparable illnesses better than frequently employed echocardiography factors. The most recent development in AI for echocardiography is a video-based deep learning (DL) system that outperforms human specialists in tasks like left ventricle segmentation, EF calculation, and the evaluation of cardiomyopathies. This algorithm's forecasts have a variance that is equal to or lower than observations of cardiac activity made by qualified humans. The first video-based DL model for echocardiography, Econet-Dynamic, predicts EF and the prevalence of heart failure with decreased EF using spatiotemporal convolutions with residual connections. This algorithm can accurately and quickly diagnose cardiovascular illness by detecting even the smallest variations in EF.

8.6.3 SINGLE-PHOTON EMISSION COMPUTED TOMOGRAPHY

Myocardial perfusion and the prevalence of blocked coronary artery disease can both be evaluated using single-photon emission computed tomography (SPECT) stress testing (D'Antonio et al., 2023). Improvements in picture capture, reconstruction, and automatic quantitation have also been made as a result of advances in AI. Cardiac magnetic resonance (CMR) imaging is another diagnostic technique used for non-invasive evaluation of cardiovascular disease (CVD). CMR (Cardiovascular Magnetic Resonance) is frequently used in the detection of numerous CVDs including cardiomyopathies, valvular heart disease, and ischemic heart disease. It is regarded as the gold standard for assessing cardiac shape, function, perfusion, and myocardial tissue measurement. However, it can take some time to gather high-quality pictures with various contrasts and complete cardiac covering. ML integration into CMR can result in a more effective scanning procedure and precise outcome analysis. A pre-learned model is used in DL based MRI reconstruction to streamline the rebuilding process for fresh data in the area of CMR (Elad, Figueiredo, & Ma, 2010).

8.6.4 HEART FAILURE

Heart failure (HF) is a prevalent condition affecting a significant proportion of adults in developed countries, and early detection and prediction of HF are crucial for effective treatment and improved patient outcomes (Lv & Zhang, 2019). ML has emerged as a promising tool for identifying HF early, classifying its severity, and predicting adverse events, such as hospital readmissions within 30 days. In addition, cardiac resynchronization therapy (CRT) is an important treatment for symptomatic HF patients with reduced ejection fraction and wide QRS complex due to intraventricular conduction delay. In general, patients who meet the criteria of having left bundle branch block (LBBB) and a QRS duration of 150 MS or more are considered eligible for cardiac resynchronization therapy (CRT) implantation. This patient population typically experiences significant reductions in mortality and readmissions following CRT. However, it is important to note that approximately 30% of patients who meet these criteria and receive a CRT implant do not clinically benefit from the therapy. Hence, it becomes crucial to predict the potential outcome of CRT for each patient as part of the pre-implantation decision-making process.

8.7 XAI NEUROLOGICAL DISORDERS

XAI (explainable artificial intelligence) is a field of AI that focuses on creating models that can provide transparent explanations for their decisions and predictions (Lötsch, Kringel, & Ultsch, 2022). In the context of neurological disorders, XAI can help medical professionals better understand the underlying mechanisms of these diseases, leading to more accurate diagnoses and more effective treatments. Here are some examples of neurological disorders that can be diagnosed using XAI.

8.7.1 ALZHEIMER'S DISEASE

In order to find trends and indicators that point to the existence of Alzheimer's disease, XAI can assist in the analysis of brain imaging data. For successful therapy, faster and more precise diagnoses may result from this. XAI can aid medical workers in their understanding of the underlying processes underpinning Alzheimer's disease and in providing more precise diagnoses. Distinguishing Alzheimer's disease from other types of dementia, such as vascular dementia or Lewy body dementia, poses a significant challenge in the diagnostic process. This differentiation is a critical aspect that healthcare professionals must address when making a diagnosis. By examining patient data and spotting trends that are specific to Alzheimer's illness, XAI can be of assistance. Using brain imaging data analysis, for instance, XAI can find trends and biomarkers that suggest the existence of Alzheimer's disease (Vrahatis et al., 2023).

According to studies, individuals with Alzheimer's disease have smaller brains overall and less efficient glucose processing in specific parts of the brain. These trends can be found by analyzing brain imaging data with XAI, which enables early and more precise diagnoses. Furthermore, by detecting changes in brain volume and glucose metabolism that are suggestive of disease development, XAI can assist medical workers in tracking disease progression over time.

Analysis of genetic data is another field where XAI can be helpful in the identification of Alzheimer's disease (Vrahatis et al., 2023). Alzheimer's disease development risk has been related to specific genetic abnormalities. These variants can be found and their effects on illness risk can be examined using XAI. Medical professionals will be able to develop specialized therapy plans for patients who are at a high risk of developing Alzheimer's disease when they have a better knowledge of the fundamental processes underlying the disease.

Data from cognitive exams and patient surveys can also be analyzed using XAI. Memory loss and other cognitive deficits are common in people with Alzheimer's disease, and cognitive exams and questionnaires can identify these deficits. This data can be analyzed by XAI to spot trends of cognitive decline that are specific to Alzheimer's. This can assist doctors in making more precise diagnoses and customizing treatment strategies for each patient.

8.7.2 PARKINSON'S DISEASE

Millions of individuals worldwide are afflicted by the neurodegenerative disorder Parkinson's disease (Quinn, Ambrósio, & Alves, 2022). The disease results in a range of motor and non-motor symptoms due to the progressive degeneration of dopamine-producing neurons in the brain. Despite the fact that there is currently no cure for Parkinson's disease, early detection and treatment can help the condition advance more slowly and enhance the patient's quality of life. XAI has become a hopeful tool for Parkinson's disease diagnosis in recent years. The goal of XAI, a subfield of AI, is to develop models that can transparently justify their judgments and forecasts.

XAI can aid healthcare workers in understanding the underlying mechanisms of Parkinson's disease and spotting its early warning signs (Nazir, Dickson, & Akram, 2023). The analysis of gait and movement patterns is a potential use of XAI in Parkinson's disease. A wide range of motor symptoms, such as tremors, rigidity, and issues with balance and coordination, are present in Parkinson's disease. Numerous clinical evaluations, such as the Hoehn and Yahr scale and the Unified Parkinson's Disease Rating Scale (UPDRS), can be used to evaluate these symptoms. These tests, however, are subjective and susceptible to the experience of the examiner, the patient's disposition, and their degree of fatigue, among other variables.

Analyzing sensor data from wearable devices, such as accelerometers and gyroscopes, is one method for using XAI to identify Parkinson's disease. These gadgets can provide a constant stream of information on the patient's movement patterns and can be worn on the wrist, ankle, or waist. On the basis of this data, XAI algorithms can be taught to recognize patterns that are indicative of Parkinson's disease, such as a reduction in arm swing or an increase in gait speed variability. Using XAI to examine video footage of the patient's gait and movement patterns is an alternative method. It is possible to teach computer vision algorithms to recognize minute adjustments in the patient's arm swing and posture. Even before other motor symptoms become evident, these changes may be a sign of Parkinson's disease in its early stages. A number of imaging methods, including positron emission tomography (PET) and single-photon emission computed tomography (SPECT), can be used to identify the progressive degeneration of dopamine-producing neurons in the brain that characterizes Parkinson's disease. On the basis of this imaging data, XAI algorithms can be taught to recognize the distinctive patterns of neurodegeneration in Parkinson's disease.

8.7.3 MULTIPLE SCLEROSIS

Multiple sclerosis (MS) is a long-term autoimmune condition that damages the central nervous system and manifests as a variety of symptoms including muscular weakness, poor coordination, and cognitive decline (Koldanov, 2022). Currently, a clinical evaluation, imaging tests like an MRI, and laboratory tests to rule out other conditions that might have comparable symptoms are used to diagnose MS. XAI, which offers a more precise and effective method of identifying the illness, has recently emerged as a promising tool for diagnosing MS. When diagnosing MS, XAI can be used to examine MRI scans of the brain and spinal cord to find the distinctive tumors connected to the condition.

One of XAI's primary benefits is its ability to quickly and accurately process large amounts of data, which is especially helpful in the case of MS. With conventional diagnostic techniques, MRI images must be carefully analyzed and evaluated against predetermined diagnostic standards. Particularly when the lesions are small or subtle, this procedure can be time-consuming and error-prone. XAI, on the other hand, uses algorithms specially created to detect MS lesions to analyze MRI images much more quickly and accurately (Olatunji et al., 2023).

Another advantage of XAI is that it can offer more individualized diagnoses that take into consideration the particular traits of each patient. XAI models can find patterns and biomarkers unique to each patient by examining MRI images and other clinical data, resulting in more individualized and successful therapies. Although XAI is still a young area, there have been a number of encouraging studies showing how well it can detect MS. For instance, a research in the journal Neuroimage: Clinical discovered that a XAI algorithm had a sensitivity of 96.5% and a specificity of 96.4% for correctly identifying MS lesions in MRI images.

Another study showed that a XAI model could predict the diagnosis of MS with an accuracy of 88.5%. It was published in the Journal of Neurology, Neurosurgery, and Psychiatry. Overall, XAI has the potential to completely transform how MS is diagnosed and treated by offering a more precise and effective method of diagnosing the condition. XAI models can find patterns and biomarkers unique to each patient by examining MRI images and other clinical data, resulting in more individualized and successful therapies (Fabrizio, Termine, Caltagirone, & Sancesario, 2021).

8.7.4 Epilepsy

A neurological condition called epilepsy is defined by abnormal brain activity that results in seizures. Seizures can be mild or severe (Mauritz et al., 2022). Millions of people globally suffer from epilepsy, and proper diagnosis and treatment are crucial for controlling the condition and enhancing quality of life. Analyzing electroencephalogram (EEG) data, which records the electrical activity of the brain, is a typical strategy. To find patterns of abnormal brain activity that point to epilepsy, XAI can examine EEG data. For instance, using MRI data analysis, XAI can spot structural variations in the brain that are linked to seizures, like hippocampal sclerosis or cortical dysplasia. In order to identify patients who are more likely to acquire epilepsy, XAI can also analyze clinical data, including age of onset, family history, and other risk factors. The ability of XAI to produce transparent and interpretable models is one of the primary advantages it has for diagnosing epilepsy.

8.7.5 Huntington's Disease

A progressive and crippling neurological disorder called Huntington's disease is thought to affect 1 in 10,000 people globally (McAllister et al., 2022). It is brought on by a genetic mutation that causes the nerve cells in the brain to deteriorate, which causes uncontrollable movements, cognitive loss, and psychiatric symptoms. Huntington's disease has no known cure, but early detection and treatment can help those who are afflicted live better lives. Huntington's disease diagnosis can be greatly aided by XAI, which gives doctors a potent instrument for precise and individualized diagnoses. Huntington's disease can be difficult to diagnose because its early signs can be subtle and easily confused with those of other diseases. XAI can be used to analyze patient MRI scans and identify characteristic patterns of brain atrophy that are associated with Huntington's disease and used to analyze patient gait and movement patterns to identify early signs of the disease, which can help medical professionals intervene early and slow the progression of the disease. Models that are clear and easy to understand are one of XAI's main benefits in the detection of Huntington's disease. XAI models can offer thorough justifications for their judgments and predictions, in contrast to conventional ML models that can be cryptic and challenging to understand. This could contribute to more precise diagnoses and more successful treatments as medical workers gain a greater understanding of the fundamental mechanisms of Huntington's disease. The capacity of XAI to combine data from various sources is another benefit in the diagnosis of Huntington's disease. Due to the rarity of Huntington's disease, it can be challenging to gather enough data for precise XAI model training. Furthermore, since XAI models can only be as effective as the data they are educated on, biased or sparse datasets can result in incorrect or deceptive assessments. Even though using XAI to diagnose Huntington's disease presents challenges, its possible advantages make it a promising area for further study and growth.

8.8   XAI AUTOIMMUNE DISEASE DIAGNOSIS

Autoimmune diseases are a group of disorders where the immune system attacks healthy tissues in the body (Schmidt, 2011). These diseases can affect any part of the body and can be difficult to diagnose due to their complex and varied symptoms. XAI has the potential to revolutionize the diagnosis and treatment of autoimmune diseases by providing transparent and interpretable models that can identify patterns and biomarkers that are difficult to detect using traditional diagnostic methods. Here are some examples of autoimmune diseases that can be diagnosed using XAI.

8.8.1   RHEUMATOID ARTHRITIS (RA)

XAI can analyze clinical data, such as patient symptoms and blood tests, to identify patterns and biomarkers that indicate the presence of RA (Olatunji et al., 2023). This can lead to earlier and more accurate diagnoses, which is crucial for effective treatment. One of the main challenges in diagnosing RA is distinguishing it from other types of arthritis, such as osteoarthritis or psoriatic arthritis. XAI can help by analyzing patient data and identifying patterns that are characteristic of RA. For example, XAI can be used to analyze blood test results to identify markers of inflammation, such as elevated levels of C-reactive protein (CRP) or erythrocyte sedimentation rate (ESR). These markers are often elevated in people with RA, indicating the presence of systemic inflammation. XAI can analyze blood test results and identify these patterns, leading to earlier and more accurate diagnoses. Another area where XAI can be useful in the diagnosis of RA is in the analysis of imaging tests, such as X-rays, ultrasound, or MRI. RA can cause joint damage and deformity over time, and these changes can be detect using imaging tests. XAI can analyze imaging data and identify patterns of joint damage and inflammation that are characteristic of RA. This can help medical professionals make more accurate diagnoses and tailor treatment plans to the individual patient. XAI can also be used to analyze data from patient questionnaires and physical exams. RA can cause pain, stiffness, and limited mobility, and these symptoms can be detected using questionnaires and physical exams. XAI can analyze this data and identify patterns of symptom severity and disease activity that are characteristic of RA. This can help medical professionals make more accurate diagnoses and monitor disease progression over time. In conclusion, XAI has the potential to revolutionize the diagnosis and treatment of RA. By providing transparent and interpretable models, XAI can help medical professionals better understand the underlying mechanisms of the disease and make more accurate diagnoses (Mridha, Uddin, Shin, Khadka, & Mridha, 2023).

8.8.2 SYSTEMIC LUPUS ERYTHEMATOSUS (SLE)

XAI can be used to analyze a patient's medical history and laboratory data to identify patterns and biomarkers that are indicative of SLE (Aljameel et al., 2023). This can help medical professionals make accurate diagnoses and tailor treatments to the individual patient. SLE is a chronic autoimmune disease that affects multiple organ systems in the body. It is a complex disease with a wide range of symptoms, making it difficult to diagnose. XAI has the potential to improve the accuracy of SLE diagnosis by analyzing patient data and identifying patterns that are indicative of the disease. XAI can be used to analyze patient medical history, physical and neurological exams, laboratory tests, and imaging studies. By analyzing this data, XAI can identify patterns that are characteristic of SLE. For example, patients with SLE often have abnormal results on blood tests, such as low red blood cell counts, high levels of anti-nuclear antibodies, and elevated levels of inflammatory markers, such as C-reactive protein and erythrocyte sedimentation rate. XAI can analyze these blood test results and identify patterns that are indicative of SLE, leading to earlier and more accurate diagnoses. In addition, XAI can be used to analyze imaging studies, such as X-rays, CT scans, and MRI scans. Patients with SLE may have abnormalities on imaging studies, such as inflammation in the joints or organs. XAI can analyze these imaging studies and identify patterns that are characteristic of SLE, leading to earlier diagnosis and more effective treatment. Furthermore, XAI can be used to analyze patient symptoms and medical history to identify patterns that are indicative of SLE. SLE is characterized by a wide range of symptoms, including joint pain and stiffness, fatigue, skin rash, fever, and kidney problems. XAI can analyze these symptoms and identify patterns that are indicative of SLE, leading to more accurate diagnosis and more effective treatment (Marro, Molinet, Cabrio, & Villata, 2023).

8.8.3 INFLAMMATORY BOWEL DISEASE (IBD)

XAI can be used to analyze patient data, such as medical history and laboratory tests, to identify patterns and biomarkers that are indicative of IBD (Katrakazas, Ballas, Anisetti, & Spais, 2022). This can help medical professionals make accurate diagnoses and develop personalized treatment plans. IBD is a chronic condition that affects the digestive system, causing inflammation and damage to the gastrointestinal tract. The two main types of IBD are Crohn's disease and ulcerative colitis, both of which can cause a range of symptoms, including abdominal pain, diarrhea, and rectal bleeding. Diagnosing IBD can be challenging, and it often requires a combination of medical history, physical and laboratory exams, and imaging tests. However, with the emergence of XAI, there is hope that diagnosis and treatment of IBD will become more accurate and effective. XAI is an emerging field of AI that focuses on creating models that can provide transparent explanations for their decisions and predictions. In the context of IBD, XAI can help medical professionals better understand the underlying mechanisms of the disease and make more accurate diagnoses. One of the main challenges in diagnosing IBD is distinguishing it from other gastrointestinal conditions, such as IBS or infectious diarrhea. XAI can help by analyzing patient data and identifying patterns that are characteristic of IBD. For example, XAI can be used to analyze data from imaging tests, such as CT scans, MRI scans, or ultrasound to identify patterns of inflammation and damage in the gastrointestinal tract. Studies have shown that people with IBD have distinct patterns of inflammation and damage in the gastrointestinal tract that are different from those seen in other gastrointestinal conditions. XAI can analyze imaging data and identify these patterns, leading to more accurate diagnoses and more personalized treatment plans. Another area where XAI can be useful in the diagnosis of IBD is in the analysis of genetic data. Certain genetic mutations have been linked to an increased risk of developing IBD. XAI can help identify these mutations and analyze their impact on disease risk. This can help medical professionals better understand the underlying mechanisms of the disease and develop personalized treatment plans for patients with a high risk of developing IBD. XAI can also be used to analyze data from laboratory tests, such as blood tests or stool samples (Thimoteo et al., 2022). People with IBD often have abnormal levels of certain markers in their blood or stool, such as CRP or fecal calprotectin. XAI can analyze this data and identify patterns of abnormal markers that are characteristic of IBD. This can help medical professionals make more accurate diagnoses and monitor disease progression over time.

8.8.4 PSORIASIS

XAI can analyze patient data, such as medical history and laboratory tests, to identify patterns and biomarkers that indicate the presence of psoriasis (Todke & Shah, 2018). Psoriasis is a persistent autoimmune skin disorder impacting a considerable number of individuals globally. This condition manifests as reddish, flaky patches on the skin, which often provoke discomfort and itching. Diagnosing psoriasis can be challenging, and it often requires a combination of physical examination, medical history, and biopsy. However, with the emergence of XAI, there is hope that diagnosis and treatment of psoriasis will become more effective. XAI is an emerging field of AI that focuses on creating models that can provide transparent explanations for their decisions and predictions. In the context of psoriasis, XAI can help medical professionals better understand the underlying mechanisms of the disease and make more accurate diagnoses. One of the main challenges in diagnosing psoriasis is distinguishing it from other skin conditions, such as eczema or rosacea. XAI can help by analyzing patient data and identifying patterns that are characteristic of psoriasis. For example, XAI can be used to analyze images of skin lesions and identify patterns that are characteristic of psoriasis. Psoriasis lesions have a unique appearance, with thick, silvery scales and a red base. XAI can analyze images of skin lesions and identify these patterns, leading to earlier and more accurate diagnoses.

Furthermore, XAI can help medical professionals monitor disease progression over time by identifying changes in lesion appearance that are indicative of disease progression. Another area where XAI can be useful in the diagnosis of psoriasis is in the analysis of genetic data. Certain genetic mutations have been linked to an increased risk of developing psoriasis. XAI can help identify these mutations and analyze their impact on disease risk. This can help medical professionals better understand the underlying mechanisms of the disease and develop personalized treatment plans for patients with a high risk of developing psoriasis. XAI can also be used to analyze data from patient questionnaires and other diagnostic tests. People with psoriasis may experience a range of symptoms, including itching, pain, and social isolation (Thimoteo et al., 2022). ML has the potential to transform the clinical management of autoimmune diseases by identifying patterns within patient data and predicting patient outcomes. A systematic review of literature was conducted to explore the use of ML in addressing clinical problems related to autoimmune diseases. The search included studies with "ML" or "AI" and search terms related to autoimmune diseases. Of the 702 studies found, 169 met the inclusion criteria. Support vector machines and random forests were the ML methods commonly utilized in various studies. Among autoimmune diseases, multiple sclerosis, rheumatoid arthritis, and inflammatory bowel disease received substantial attention in research. However, it is worth noting that only a limited number of studies incorporated diverse data types during the modeling process and employed cross-validation and separate testing sets to ensure the robustness of model evaluation. By embracing these best practices, such as rigorous validation and independent testing of ML models, it is possible to develop more precise and dependable models. While many models demonstrated favorable predictive outcomes in simple scenarios, the integration of multiple data types holds the potential for the development of more intricate predictive models in the future.

8.9 CASE STUDIES OF XAI APPLICATIONS IN DISEASE DIAGNOSIS

There are several case studies showcasing the applications of XAI in disease diagnosis. Here are a few examples.

8.9.1 SKIN CANCER DIAGNOSIS

Skin cancer is a prevalent type of cancer that has a significant impact on numerous individuals globally. Conventionally, the diagnosis of skin cancer relies on dermatologists who visually inspect the skin lesion and determine its malignancy. Nevertheless, human errors can occur, leading to delayed or unnecessary treatments. To tackle this issue, Google and Stanford University joined forces to conduct a study assessing the effectiveness of a deep learning algorithm in skin cancer diagnosis (Esteva et al., 2017). The algorithm underwent training using a vast dataset comprising more than 130,000 images of skin lesions obtained from diverse sources (Lee, Kim, Jeong, Choi, & science, 2018). The algorithm was designed to classify the images as either cancerous or non-cancerous, based on a set of predetermined criteria. After training the algorithm, the researchers tested its accuracy by comparing its diagnosis with that of a group of dermatologists. The results of the study showed that the algorithm was able to accurately identify cancerous lesions with an accuracy of 91% (Kallaway et al., 2013). The algorithm also outperformed the dermatologists, who had an accuracy of 86% (Rezvantalab, Safigholi, & Karimijeshni, 2018). The researchers also noted that the algorithm was able to diagnose a broader range of skin lesions than the dermatologists, which suggests that it has the potential to improve the accuracy of skin cancer diagnosis. To further improve the trustworthiness of the algorithm, the researchers used an XAI model to provide a visual explanation of the model's decision-making process. The XAI model used a technique called "attention mapping" to highlight the parts of the image the algorithm used to make its diagnosis. This allowed physicians to see how the algorithm arrived at its conclusion and provided them with more confidence in the results.

8.9.2 Pneumonia Diagnosis

Pneumonia is a potentially life-threatening disease that affects millions of people every year (Bhadra, Gupta, Garg, Dhangar, & Jadhav, 2023). Diagnosis of pneumonia typically involves a chest X-ray, which can be time-consuming for physicians to interpret and may lead to errors in diagnosis. To address this challenge, researchers at Stanford University developed an XAI algorithm that could accurately diagnose pneumonia from chest X-rays. The XAI algorithm used a convolutional neural network to analyze chest X-rays and identify patterns associated with pneumonia. The algorithm was trained on a large dataset of chest X-rays and pneumonia diagnoses, allowing it to learn the key features of pneumonia on X-rays. One of the unique features of this XAI algorithm is the provision of heat maps that highlight the regions of the X-ray that were most indicative of pneumonia. This feature allowed physicians to better understand the model's decision-making process and increased their confidence in the results. The XAI algorithm was evaluated on a dataset of over 100,000 chest X-rays and achieved an accuracy of 92% in diagnosing pneumonia. This level of accuracy is comparable to that of human radiologists and highlights the potential of XAI in improving medical diagnosis.

8.9.3 Prostate Cancer Diagnosis

The use of XAI in disease diagnosis has been gaining momentum, and the researchers at the University of Texas have demonstrated its potential in prostate cancer detection (Qian, Li, Wang, & He, 2023). By leveraging MRI images, they developed an algorithm that was able to predict the presence of cancer in the prostate. The XAI model used a heatmap visualization technique that highlighted the areas of the prostate that were most indicative of cancer, providing a more transparent and interpretable output. The results of the study were impressive, with the algorithm achieving an accuracy of 89%. This is a significant improvement over traditional diagnostic methods, which can have lower accuracy rates and are often subjective. By using XAI, the researchers were able to provide a more reliable and trustworthy method for prostate cancer detection, which could ultimately lead to better patient outcomes. The XAI model developed by the researchers has the potential to be implemented in clinical settings, providing doctors with a more accurate and efficient tool for prostate cancer diagnosis. Furthermore, the XAI model's explainability allows for transparency and interpretability, which can lead to increased trust and confidence in the algorithm's predictions. Overall, this study highlights the potential for XAI to revolutionize disease diagnosis and improve patient outcomes.

8.9.4 Diabetic Retinopathy Diagnosis

XAI has the potential to revolutionize disease diagnosis and improve patient outcomes. The use of ML algorithms in healthcare is rapidly increasing, and XAI models are becoming more popular due to their ability to provide interpretable results. In the case of diabetic retinopathy, the XAI algorithm developed by Google allowed physicians to understand the decision-making process of the model, which helped to build trust in the results. Another example of the use of XAI in disease diagnosis is in the field of cardiology. Researchers at the University of Michigan used an XAI model to predict which patients were at high risk of developing heart failure. The model used electronic health records, including demographic information, medical history, laboratory tests, and medications, to predict heart failure. The XAI model provided an interpretable risk score that allowed physicians to understand the factors that contributed to the risk prediction. XAI models have also been used in cancer diagnosis. Researchers at the University of Texas used an XAI model to predict which patients with lung cancer were likely to respond to immunotherapy. The model used genomic data to predict the likelihood of response to treatment, and the XAI model provided an interpretable score that allowed physicians to understand the factors that contributed to the prediction. In each of these instances, XAI models demonstrated their ability to generate interpretable outcomes, aiding physicians in comprehending the decision-making process of the algorithm (Ali et al., 2023). This heightened transparency bolstered trust in the model and empowered physicians to make better-informed decisions. The potential of XAI to enhance the accuracy and dependability of disease diagnosis holds promise for improved patient outcomes and the development of personalized treatment approaches.

8.10 FUTURE DIRECTIONS AND CHALLENGES

8.10.1 Emerging Trends in XAI for Disease Diagnosis

AI possesses immense potential to revolutionize various aspects of medicine, both presently and in the future (Agrawal, Nikhade, & Nikhade, 2022). However, the lack of transparency in AI applications has emerged as a significant concern, not only within the medical field but across different domains. This issue becomes particularly critical when users must interpret the outcomes generated by AI systems. XAI offers a solution by providing a rationale that enables users to comprehend the reasoning behind a system's output. This interpretability allows for context-specific interpretation of the output (Agrawal et al., 2022). Clinical

Decision Support Systems (CDSSs) are an area that urgently requires XAI implementation. These systems assist medical practitioners in their decision-making processes, and without explain ability, there is a risk of both under-reliance and over-reliance on their recommendations. By offering explanations for the basis of recommendations, practitioners can make more nuanced and potentially life-saving decisions. The necessity for XAI in CDSSs, as well as the wider medical field, is further amplified by the importance of ethical and fair decision-making. It is crucial to uncover and address any historical actions and biases embedded in AI trained on historical data

The advancement of XAI has been propelled by various factors, including the development of novel statistical learning algorithms, the availability of large datasets, and the accessibility of cost-effective yet powerful hardware, such as cloud storage (Albahri et al., 2023). These advancements have made XAI tools widely feasible and increasingly prevalent. The application of XAI is expanding beyond software engineering into socially sensitive domains like education, law enforcement, forensics, and healthcare. However, the use of XAI in such domains presents complexities, particularly when the inner workings of these systems are beyond the understanding of those affected by their predictions (Javed et al., 2023).

The medical field presents numerous challenges that can potentially be addressed through the implementation of AI. Significant progress has been made in automated diagnosis, prognosis, drug design, and testing, driven by the importance of medical care and the abundance of data from sources like medical imaging, biosensors, molecular data, and electronic medical records. The goals of AI in medicine include personalized medical decisions, health practices, and therapies tailored to individual patients. Nevertheless, the current state of AI in medicine is described as having high potential but limited data and proof. While some AI-based systems have been validated in real-world settings for various tasks such as diabetic retinopathy detection, wrist fracture detection, histologic breast cancer metastases identification, detection of very small colonic polyps, and congenital cataracts, many of these systems have demonstrated high false-positive rates when deployed in real-world clinical environments. It is important to note some emerging trends in XAI for disease diagnosis.

8.10.1.1 Integration of XAI with Clinical Decision-making

XAI is being integrated into clinical decision-making tools to help healthcare professionals make more accurate and transparent decisions (Al-Antari, 2023). This involves using XAI to explain how a particular diagnosis was made by a ML model, and allowing healthcare professionals to modify the model's output based on their own clinical judgment.

8.10.1.1.1 Development of Hybrid Models

Hybrid models that combine ML with expert knowledge are becoming increasingly popular for disease diagnosis (Al Bataineh & Manacek, 2022). These models use XAI to explain how they arrived at a particular diagnosis, making it easier for healthcare professionals and patients to understand and trust the model's output.

8.10.1.1.2 Explain Ability for Deep Learning Models

Deep learning models, which are highly complex and difficult to interpret, are being made more explainable using XAI techniques such as attention maps and saliency maps (Mankodiya et al., 2022). These techniques highlight the areas of an image or dataset the model is focusing on when making a diagnosis, making it easier to understand how the model arrived at its conclusion.

8.10.1.1.3 Visualization tools for XAI

Visualization tools are being developed to help healthcare professionals and patients better understand how XAI models make diagnoses (Dey et al., 2022). These tools use interactive graphics and visualizations to show how the model is processing data and making predictions, making it easier to understand and trust the model's output.

Overall, XAI has the potential to greatly improve disease diagnosis by making ML models more transparent and interpretable. As the field continues to develop, we can expect to see more innovative applications of XAI in healthcare.

8.11 CHALLENGES AND LIMITATIONS OF XAI IN DISEASE DIAGNOSIS

XAI in medicine faces several challenges and limitations that encompass areas such as bias, privacy, security, transparency, causality, transferability, informativeness, fairness, and confidence. These concerns are especially critical because decisions influenced by AI systems have a direct impact on human health. Therefore, there is a pressing need to comprehend how these decisions are made. This need is particularly amplified in certain areas, such as disease diagnosis, where the outcomes and decisions can be life-changing (Park et al., 2022). Additionally, in precision medicine, experts require a more comprehensive understanding of the model's outputs beyond simple binary predictions to support their diagnostic process. Overcoming these challenges is crucial to ensure the responsible and effective use of XAI in the medical field.

The use of XAI in disease diagnosis poses several many other challenges and limitations. Some of which include the following.

8.11.1 LIMITED INTERPRETABILITY OF BLACK-BOX MODELS

This makes it challenging to explain how a diagnosis was arrived at, making it difficult for medical professionals to trust the diagnosis and provide an explanation to the patient (Anestis, Eccles, Fletcher, Triliva, & Simpson, 2022).

8.11.2 LIMITED DATA AVAILABILITY

The availability of quality healthcare data is limited due to privacy concerns, data silos, and legal restrictions. This hampers the development of robust AI models that can aid in accurate diagnosis (Rahman et al., 2022).

8.11.3 LIMITED GENERALIZABILITY

ML models are trained on specific datasets, which may not be representative of the general population (J. Yang, Soltan, & Clifton, 2022). This can result in a lack of generalizability when applied to new patients, which can lead to inaccurate or biased diagnoses.

8.11.4 LACK OF TRANSPARENCY

Many AI models used in disease diagnosis are proprietary, and the algorithms used are not open for scrutiny. This can lead to a lack of transparency and trust, which is critical in the medical field.

8.11.5 ETHICAL CONCERNS

The use of AI in disease diagnosis raises ethical concerns, including privacy, bias, and the potential for harm to patients. This requires careful consideration and management to ensure that the benefits of AI in disease diagnosis outweigh the potential risks.

8.11.6 INTEGRATION WITH EXISTING HEALTHCARE SYSTEMS

Integrating AI models into existing healthcare systems can be challenging (Mbunge & Muchemwa, 2022). This requires changes in infrastructure, workflows, and staff training to ensure that AI models are effectively used and integrated into the existing healthcare system.

8.11.7 COMPLEXITY

Machine learning models can be highly complex and difficult to explain, which makes it challenging to provide a clear and concise explanation of how they work.

8.11.8 INTERPRETABILITY

The outputs of ML models can be difficult to interpret, especially when dealing with complex datasets or models.

8.11.9 PRIVACY

There are concerns about the privacy implications of XAI, as explanations of ML models may reveal sensitive information about individuals or organizations.

8.11.10 COST

Developing and implementing XAI can be costly and time-consuming, especially for large-scale applications.

Overall, the challenges and limitations of XAI in disease diagnosis highlight the need for careful consideration and management to ensure that the benefits of AI are maximized, while the risks and limitations are minimized.

8.12 POSSIBLE SOLUTIONS TO OVERCOME CHALLENGES

To overcome these challenges, the following solutions can be considered.

8.12.1 DEVELOP SIMPLER AND MORE INTERPRETABLE MODELS

One possible solution is to develop simpler and more interpretable models, such as decision trees or rule-based systems. These models are easier to understand and explain, and can be more transparent than more complex models.

8.12.2 USE VISUALIZATIONS

Visualizations can be used to help explain the outputs of ML models, making them more accessible to non-experts (Yang, Suh, Chen, & Ramos, 2018).

8.12.3 ENSURE PRIVACY

To address privacy concerns, XAI systems should be designed to protect sensitive information. One approach is to use anonymization techniques or to provide explanations that do not reveal sensitive information.

8.12.4 COLLABORATIVE DEVELOPMENT

Collaboration between experts in ML, ethics, and law can help ensure that XAI systems are developed in a way that is ethical, transparent, and legally compliant.

8.12.5 DEVELOP STANDARDS AND REGULATIONS

Developing standards and regulations for XAI can help ensure that it is used responsibly and ethically, and that it meets the needs of stakeholders such as users and regulators (Díaz-Rodríguez et al., 2023).

8.12.6 ENHANCE EDUCATION AND COMMUNICATION

Improved education and communication around XAI can help increase understanding and acceptance of XAI, and address any misconceptions or concerns (Javed et al., 2023).

8.13 CONCLUSION

In conclusion, this chapter presented a comprehensive review and survey of the existing literature on medical applications of XAI in the fields of diagnosis and surgery.

The study provides a summary of its findings, discusses the current limitations of medical XAI applications, and outlines future perspectives in this area. The interdisciplinary nature of AI and medicine is addressed, emphasizing that this review can bridge the gap between AI and medical professionals, offering valuable insights for future researchers aiming to design effective medical XAI applications. Importantly, the utilization of XAI not only provides explanations for outputs but also instills trust and reliability in end users, including doctors, thus facilitating the validation of decisions. This has the potential to benefit a wide range of individuals and enhance medical practice.

XAI can significantly enhance the accuracy and interpretability of disease diagnosis. With the ability to explain its decision-making process, XAI can help healthcare professionals and patients better understand the underlying factors that contribute to a diagnosis. This transparency can also help to build trust in the diagnosis and the technology behind it. Additionally, XAI can facilitate the development of personalized treatment plans by identifying the specific factors that led to the diagnosis. Overall, the integration of XAI in disease diagnosis has the potential to improve patient outcomes and provide a more comprehensive understanding of the factors that contribute to the development of diseases. However, continued research is needed to further develop and refine XAI models for clinical use.

REFERENCES

Agarwal, P., Swami, S., & Malhotra, S. K. (2022). Artificial intelligence adoption in the post COVID-19 new-normal and role of smart technologies in transforming business: a review. *Journal of Science and Technology Policy Management, 12*(4).

Agrawal, P., Nikhade, P., & Nikhade, P. P. (2022). Artificial intelligence in dentistry: past, present, and future. *Cureus, 14*(7), E27405-e27405.

Ahmed, Z., Mohamed, K., Zeeshan, S., & Dong, X. (2020). Artificial intelligence with multifunctional machine learning platform development for better healthcare and precision medicine. Database, 2020, baaa010.

Ahsan, M. M., Luna, S. A., & Siddique, Z. (2022). Machine-learning-based disease diagnosis: a comprehensive review. Paper Presented at the Healthcare. (Vol. 10, No. 3, p. 541). MDPI.

Al-Antari, M. A. (2023). Artificial intelligence for medical diagnostics existing and future AI technology! *Diagnostics, 13*(4), 688, https://doi.org/10.3390/diagnostics13040688

Al Bataineh, A., & Manacek, S. (2022). MLP-PSO hybrid algorithm for heart disease prediction. *Journal of Personalized Medicine, 12*(8), 1208.

Albahri, A., Duhaim, A. M., Fadhel, M. A., Alnoor, A., Baqer, N. S., Alzubaidi, L., ... Salhi, A. (2023). A systematic review of trustworthy and explainable artificial intelligence in healthcare: assessment of quality, bias risk, and data fusion. Information Fusion, 96(10), 156–191.

Ali, S., Abuhmed, T., El-Sappagh, S., Muhammad, K., Alonso-Moral, J. M., Confalonieri, R., ... Herrera, F. (2023). Explainable Artificial Intelligence (XAI): what we know and what is left to attain Trustworthy Artificial Intelligence. *Information Fusion*, 101805.

Aljameel, S. S., Alzahrani, M., Almusharraf, R., Altukhais, M., Alshaia, S., Sahlouli, H., ... Alsumayt, A. (2023). Prediction of preeclampsia using machine learning and deep learning models: a review. *Big Data and Cognitive Computing, 7*(1), 32.

Anestis, E., Eccles, F. J., Fletcher, I., Triliva, S., & Simpson, J. (2022). Healthcare professionals' involvement in breaking bad news to newly diagnosed patients with motor neurodegenerative conditions: a qualitative study. *Disability and Rehabilitation, 44*(25), 7877–7890.

Antoniadi, A. M., Du, Y., Guendouz, Y., Wei, L., Mazo, C., Becker, B. A., & Mooney, C. (2021). Current challenges and future opportunities for XAI in machine learning-based clinical decision support systems: a systematic review. *Applied Sciences, 11*(11), 5088.

Anvari, M. S., Gharib, A., Abolhasani, M., Azari-Yam, A., Gharalari, F. H., Safavi, M., ... Vasei, M. (2021). Pre-analytical practices in the molecular diagnostic tests, a concise review. *Iranian Journal of Pathology, 16*(1), 1.

Arrieta, A. B., Díaz-Rodríguez, N., Del Ser, J., Bennetot, A., Tabik, S., Barbado, A., ... Benjamins, R. (2020). Explainable Artificial Intelligence (XAI): concepts, taxonomies, opportunities and challenges toward responsible AI. *Information Fusion, 58*, 82–115.

Benjamins, J.-W., Yeung, M. W., Reyes-Quintero, A. E., Ruijsink, B., van der Harst, P., & Juarez-Orozco, L. E. (2022). Hybrid cardiac imaging: the role of machine learning and artificial intelligence. In Francesco Nudi, Orazio Schillaci, Giuseppe Biondi-Zoccai, Ami E. Iskandrian (eds.) *Hybrid Cardiac Imaging for Clinical Decision-Making: From Diagnosis to Prognosis* (pp. 203–222). Springer, Nature Switzerland AG.

Bhadra, A., Gupta, S., Garg, N., Dhangar, S., & Jadhav, S. (2023). Survey on pneumonia disease detection. *International Journal of Engineering Research & Technology (IJERT), 12*(04). www.ijert.org

Bodenheimer, T. (2008). Coordinating care-a perilous journey through the health care system. *New England Journal of Medicine, 358*(10), 1064.

Burrell, J. (2016). How the machine 'thinks': understanding opacity in machine learning algorithms. *Big Data & Society, 3*(1), 2053951715622512.

Chattu, V. K. (2021). A review of artificial intelligence, big data, and blockchain technology applications in medicine and global health. *Big Data and Cognitive Computing, 5*(3), 41.

D'Antonio, A., Assante, R., Zampella, E., Mannarino, T., Buongiorno, P., Cuocolo, A., & Acampa, W. (2023). Myocardial blood flow evaluation with dynamic cadmium-zinc-telluride single-photon emission computed tomography: bright and dark sides. Diagnostic and Interventional Imaging, *104*(7–8), S2211–5684 (23), 00025.

Dağlarli, E. (2020). Explainable artificial intelligence (xAI) approaches and deep meta-learning models. *Advances and Applications in Deep Learning, 79*.

de-Lima-Santos, M.-F., & Ceron, W. (2021). Artificial intelligence in news media: current perceptions and future outlook. *Journalism and Media, 3*(1), 13–26.

De Bruijne, M. J. M. i. a. (2016). Machine learning approaches in medical image analysis: From detection to diagnosis. *Medical Image Analysis, 100*(33), 94–97.

Dees, N. D., Zhang, Q., Kandoth, C., Wendl, M. C., Schierding, W., Koboldt, D. C., ... Mardis, E. R. (2012). MuSiC: identifying mutational significance in cancer genomes. *Genome Research, 22*(8), 1589–1598.

Dell'Angela, L., & Nicolosi, G. L. (2022). Artificial intelligence applied to cardiovascular imaging, a critical focus on echocardiography: the point-of-view from "the other side of the coin". *Journal of Clinical Ultrasound, 50*(6), 772–780.

Della Bella, P., Baratto, F., Vergara, P., Bertocchi, P., Santamaria, M., Notarstefano, P., ... Piacenti, M. (2022). Does timing of ventricular tachycardia ablation affect prognosis in patients with an implantable cardioverter defibrillator? Results from the multicenter randomized PARTITA trial. *Circulation, 145*(25), 1829–1838.

Dey, S., Chakraborty, P., Kwon, B. C., Dhurandhar, A., Ghalwash, M., Saiz, F. J. S., ... Meyer, P. (2022). Human-centered explainability for life sciences, healthcare, and medical informatics. *Patterns, 3*(5), 100493.

Díaz-Rodríguez, N., Del Ser, J., Coeckelbergh, M., de Prado, M. L., Herrera-Viedma, E., & Herrera, F. (2023). Connecting the dots in trustworthy artificial intelligence: from AI principles, ethics, and key requirements to responsible AI systems and regulation. arXiv preprint *arXiv:2305.02231*.

Du, M., Liu, N., & Hu, X. (2019). Techniques for interpretable machine learning. *Communications of the ACM, 63*(1), 68–77.

Elad, M., Figueiredo, M. A., & Ma, Y. (2010). On the role of sparse and redundant representations in image processing. *Proceedings of the IEEE, 98*(6), 972–982.

Esteva, A., Kuprel, B., Novoa, R. A., Ko, J., Swetter, S. M., Blau, H. M., & Thrun, S. J. n. (2017). Dermatologist-level classification of skin cancer with deep neural networks. Nature, *542*(7639), 115–118.

Fabrizio, C., Termine, A., Caltagirone, C., & Sancesario, G. (2021). Artificial intelligence for Alzheimer's disease: promise or challenge? *Diagnostics, 11*(8), 1473.

Ghazal, T. M., Hasan, M. K., Alshurideh, M. T., Alzoubi, H. M., Ahmad, M., Akbar, S. S., … Akour, I. A. (2021). IoT for smart cities: machine learning approaches in smart healthcare—A review. *Future Internet, 13*(8), 218.

Golden, J. A. J. J. (2017). Deep learning algorithms for detection of lymph node metastases from breast cancer: helping artificial intelligence be seen. *JAMA, 318*(22), 2184–2186.

Goldenberg, S. L., Nir, G., & Salcudean, S. E. (2019). A new era: artificial intelligence and machine learning in prostate cancer. *Nature Reviews Urology, 16*(7), 391–403.

Groves, P., Kayyali, B., Knott, D., & Kuiken, S. V. (2016). *The 'big data' revolution in healthcare: accelerating value and innovation.* Scientific and Technological information Resources, Colombia.

Haider, A., Ringer, M., Kotroczó, Z., Mohácsi-Farkas, C., & Kocsis, T. (2023). The current level of MALDI-TOF MS applications in the detection of microorganisms: a short review of benefits and limitations. *Microbiology Research, 14*(1), 80–90.

Hashimoto, D. A., Rosman, G., Rus, D., & Meireles, O. R. (2018). Artificial intelligence in surgery: promises and perils. *Annals of Surgery, 268*(1), 70–76.

Hou, T.-Y., Chiang-Ni, C., & Teng, S.-H. (2019). Current status of MALDI-TOF mass spectrometry in clinical microbiology. *Journal of Food and Drug Analysis, 27*(2), 404–414.

Hussain, H. K., Tariq, A., & Gill, A. Y. (2023). Role of AI in cardiovascular health care; a brief overview. *Journal of World Science, 2*(4), 794–802.

Islam, M. R., Ahmed, M. U., Barua, S., & Begum, S. (2022). A systematic review of explainable artificial intelligence in terms of different application domains and tasks. *Applied Sciences, 12*(3), 1353.

Jain, S., Nehra, M., Kumar, R., Dilbaghi, N., Hu, T., Kumar, S., … Li, C.-Z. (2021). Internet of medical things (IoMT)-integrated biosensors for point-of-care testing of infectious diseases. *Biosensors and Bioelectronics, 179*, 113074.

Javed, A. R., Ahmed, W., Pandya, S., Maddikunta, P. K. R., Alazab, M., & Gadekallu, T. R. (2023). A survey of explainable artificial intelligence for smart cities. *Electronics, 12*(4), 1020.

Jaycox, J. R., Lucas, C., Yildirim, I., Dai, Y., Wang, E. Y., Monteiro, V., … Buckner, J. H. (2023). SARS-CoV-2 mRNA vaccines decouple anti-viral immunity from humoral autoimmunity. *Nature Communications, 14*(1), 1299.

Kallaway, C., Almond, L. M., Barr, H., Wood, J., Hutchings, J., Kendall, C., … therapy, p. (2013). Advances in the clinical application of Raman spectroscopy for cancer diagnostics. Photodiagnosis and Photodynamic Therapy, *10*(3), 207–219.

Katrakazas, P., Ballas, A., Anisetti, M., & Spais, I. (2022). An artificial intelligence outlook for colorectal cancer screening. *arXiv preprint arXiv:2209.12624*.

Koldanov, N. (2022). *Early typical and atypical multiple sclerosis (MS) symptoms*. Lietuvos sveikatos mokslų universitetas, Lithuanian University of Health Sciences, Faculty of medicine Department of Neurology.

Krittanawong, C., Zhang, H., Wang, Z., Aydar, M., & Kitai, T. (2017). Artificial intelligence in precision cardiovascular medicine. *Journal of the American College of Cardiology, 69*(21), 2657–2664.

Lagathu, G., Grolhier, C., Besombes, J., Maillard, A., Comacle, P., Pronier, C., & Thibault, V. (2023). Using discarded facial tissues to monitor and diagnose viral respiratory infections. *Emerging Infectious Diseases, 29*(3).

Lee, J.-H., Kim, D.-h., Jeong, S.-N., Choi, S.-H. J. J. o. p., & science, i. (2018). Diagnosis and prediction of periodontally compromised teeth using a deep learning-based convolutional neural network algorithm. Journal of Periodontal & Implant Science, *48*(2), 114–123.

Little, S. H., Rigolin, V. H., Garcia-Sayan, E., Hahn, R. T., Hung, J., Mackensen, G. B., … Saric, M. (2023). Recommendations for special competency in echocardiographic guidance of structural heart disease interventions: from the American society of echocardiography. *Journal of the American Society of Echocardiography, 36*(4), 350–365.

Loomba, K., Shi, D., Sherpa, T., Chen, J., Daniels, T. J., Pavia, C. S., & Zhang, D. (2023). Use of the Western blot technique to identify the immunogenic proteins of Borrelia burgdorferi for developing a Lyme disease vaccine. *Biomedicine & Pharmacotherapy, 157*, 114013.

Lötsch, J., Kringel, D., & Ultsch, A. (2022). Explainable artificial intelligence (XAI) in biomedicine: making AI decisions trustworthy for physicians and patients. *BioMedInformatics, 2*(1), 1–17.

Lu, J., Hutchens, R., Hung, J., Bennamoun, M., McQuillan, B., Briffa, T., … Chow, B. (2022). Performance of multilabel machine learning models and risk stratification schemas for predicting stroke and bleeding risk in patients with non-valvular atrial fibrillation. *Computers in Biology and Medicine, 150*, 106126.

Lu, Y.-C., Xiao, Y., Sears, A., & Jacko, J. A. (2005). A review and a framework of handheld computer adoption in healthcare. *International Journal of Medical Informatics, 74*(5), 409–422.

Lv, J.-C., & Zhang, L.-X. (2019). Prevalence and disease burden of chronic kidney disease. *Renal Fibrosis: Mechanisms and Therapies, 1165*, 3–15.

Majnarić, L. T., Babič, F., O'Sullivan, S., & Holzinger, A. (2021). AI and big data in healthcare: towards a more comprehensive research framework for multimorbidity. *Journal of Clinical Medicine, 10*(4), 766.

Mankodiya, H., Jadav, D., Gupta, R., Tanwar, S., Hong, W.-C., & Sharma, R. (2022). Od-XAI: explainable AI-based semantic object detection for autonomous vehicles. *Applied Sciences, 12*(11), 5310.

Marro, S., Molinet, B., Cabrio, E., & Villata, S. (2023, Feb). *Natural Language Explanatory Arguments for Correct and Incorrect Diagnoses of Clinical Cases.* Paper Presented at the ICAART 2023-15th International Conference on Agents and Artificial Intelligence, Lisbon (Portugal).

Mauritz, M., Hirsch, L. J., Camfield, P., Chin, R., Nardone, R., Lattanzi, S., & Trinka, E. (2022). Acute symptomatic seizures: an educational, evidence-based review. *Epileptic Disorders, 24*(1), 26–49.

Mbunge, E., & Muchemwa, B. (2022). Towards emotive sensory Web in virtual health care: trends, technologies, challenges and ethical issues. *Sensors International, 3*, 100134.

McAllister, B., Donaldson, J., Binda, C. S., Powell, S., Chughtai, U., Edwards, G., ... Schuhmacher, L.-N. (2022). Exome sequencing of individuals with Huntington's disease implicates FAN1 nuclease activity in slowing CAG expansion and disease onset. *Nature Neuroscience, 25*(4), 446–457.

Mishra, S., Dash, A., & Jena, L. (2021). Use of deep learning for disease detection and diagnosis. *Bio-Inspired Neurocomputing, 903,* 181–201.

Mridha, K., Uddin, M. M., Shin, J., Khadka, S., & Mridha, M. (2023). An interpretable skin cancer classification using optimized convolutional neural network for a smart healthcare system. *IEEE Access, 11,* 41003–41018, Electronic ISSN: 2169-3536.

Nagura-Ikeda, M., Imai, K., Tabata, S., Miyoshi, K., Murahara, N., Mizuno, T., ... Iwata, M. (2020). Clinical evaluation of self-collected saliva by quantitative reverse transcription-PCR (RT-qPCR), direct RT-qPCR, reverse transcription–loop-mediated isothermal amplification, and a rapid antigen test to diagnose COVID-19. *Journal of Clinical Microbiology, 58*(9), e01438–01420.

Nazir, S., Dickson, D. M., & Akram, M. U. (2023). Survey of explainable artificial intelligence techniques for biomedical imaging with deep neural networks. *Computers in Biology and Medicine*, 106668.

Olatunji, S. O., Alsheikh, N., Alnajrani, L., Alanazy, A., Almusairii, M., Alshammasi, S., ... Basheer Ahmed, M. I. (2023). Comprehensible machine-learning-based models for the pre-emptive diagnosis of multiple sclerosis using clinical data: a retrospective study in the eastern province of Saudi Arabia. *International Journal of Environmental Research and Public Health, 20*(5), 4261.

Páez, A. (2019). The pragmatic turn in explainable artificial intelligence (XAI). *Minds and Machines, 29*(3), 441–459.

Park, J. S., Saeidian, A. H., Youssefian, L., Kondratuk, K. E., Pride, H. B., Vahidnezhad, H., & Uitto, J. (2022). Inherited ichthyosis as a paradigm of rare skin disorders: genomic medicine, pathogenesis, and management. *Journal of the American Academy of Dermatology, 89*(6), 1215–1226.

Patel, M., Adnan, M., Aldarhami, A., Bazaid, A. S., Saeedi, N. H., Alkayyal, A. A., ... Alshaghdali, K. (2023). Current insights into diagnosis, prevention strategies, treatment, therapeutic targets, and challenges of Monkeypox (Mpox) infections in human populations. *Life, 13*(1), 249.

Payrovnaziri, S. N., Chen, Z., Rengifo-Moreno, P., Miller, T., Bian, J., Chen, J. H., ... He, Z. (2020). Explainable artificial intelligence models using real-world electronic health record data: a systematic scoping review. *Journal of the American Medical Informatics Association, 27*(7), 1173–1185.

Prabhakararao, E., & Dandapat, S. (2022). Atrial fibrillation burden estimation using multitask deep convolutional neural network. *IEEE Journal of Biomedical and Health Informatics, 26*(12), 5992–6002.

Prieto-Avalos, G., Cruz-Ramos, N. A., Alor-Hernández, G., Sánchez-Cervantes, J. L., Rodríguez-Mazahua, L., & Guarneros-Nolasco, L. R. (2022). Wearable devices for physical monitoring of heart: a review. *Biosensors, 12*(5), 292.

Purohit, M., & Mustafa, T. (2015). Laboratory diagnosis of extra-pulmonary tuberculosis (EPTB) in resource-constrained setting: state of the art, challenges and the need. *Journal of Clinical and Diagnostic Research: JCDR, 9*(4), EE01.

Qian, J., Li, H., Wang, J., & He, L. (2023). Recent advances in explainable artificial intelligence for magnetic resonance imaging. *Diagnostics, 13*(9), 1571.

Quinn, P. M., Ambrósio, A. F., & Alves, C. H. (2022). *Oxidative Stress, Neuroinflammation and Neurodegeneration: The Chicken, the Egg and the Dinosaur.* Antioxidants, 11, 1554. MDPI, Basel, Switzerland.

Rahman, A., Hossain, M. S., Muhammad, G., Kundu, D., Debnath, T., Rahman, M., ... Band, S. S. (2022). Federated learning-based AI approaches in smart healthcare: concepts, taxonomies, challenges and open issues. *Cluster Computing, 26*(4), 1–41.

Rezvantalab, A., Safigholi, H., & Karimijeshni, S. J. a. p. a. (2018). *Dermatologist level dermoscopy skin cancer classification using different deep learning convolutional neural networks algorithms.* Cornell University.

Ribeiro, M. T., Singh, S., & Guestrin, C. (2016). Model-agnostic interpretability of machine learning. *arXiv preprint arXiv:1606.05386.*

Robbins, S. (2019). A misdirected principle with a catch: explicability for AI. *Minds and Machines, 29*(4), 495–514.

Rosado, J., Pelleau, S., Cockram, C., Merkling, S. H., Nekkab, N., Demeret, C., ... Fafi-Kremer, S. (2021). Multiplex assays for the identification of serological signatures of SARS-CoV-2 infection: an antibody-based diagnostic and machine learning study. *The Lancet Microbe, 2*(2), e60–e69.

Roy, S., Meena, T., & Lim, S.-J. (2022). Demystifying supervised learning in healthcare 4.0: a new reality of transforming diagnostic medicine. *Diagnostics, 12*(10), 2549.

Sarno, L., Neola, D., Carbone, L., Saccone, G., Carlea, A., Miceli, M., ... Di Girolamo, R. (2022). Use of artificial intelligence in obstetrics: not quite ready for prime time. American Journal of Obstetrics & Gynecology MFM, *5*(2), 100792.

Schmidt, C. W. (2011). Questions persist: environmental factors in autoimmune disease. *National Institute of Environmental Health Sciences, 119*(6), A248–A253.

Sharma, S., Henderson, J., & Ghosh, J. (2020, Feb). *CERTIFAI: A Common Framework to Provide Explanations and Analyse the Fairness and Robustness of Black-Box Models.* Paper presented at the Proceedings of the AAAI/ACM Conference on AI, Ethics, and Society, USA.

Thimoteo, L. M., Vellasco, M. M., Amaral, J., Figueiredo, K., Yokoyama, C. L., & Marques, E. (2022). Explainable artificial intelligence for COVID-19 diagnosis through blood test variables. *Journal of Control, Automation and Electrical Systems, 33*(2), 625–644.

Todke, P., & Shah, V. H. J. I. J. o. D. (2018). Psoriasis: implication to disease and therapeutic strategies, with an emphasis on drug delivery approaches. International Journal of Dermatology, *57*(11), 1387–1402.

Tran, N. K., Albahra, S., Rashidi, H., & May, L. (2022). Innovations in infectious disease testing: leveraging COVID-19 pandemic technologies for the future. Clinical Biochemistry, *117*, 10–15.

Tsai, H., Phinney, B. S., Grigorean, G., Salemi, M. R., Rashidi, H. H., Pepper, J., & Tran, N. K. (2022). Identification of endogenous peptides in nasal swab transport media used in MALDI-TOF-MS based COVID-19 screening. *ACS Omega, 7*(20), 17462–17471.

Van der Velden, B. H., Kuijf, H. J., Gilhuijs, K. G., & Viergever, M. A. (2022). Explainable artificial intelligence (XAI) in deep learning-based medical image analysis. *Medical Image Analysis, 79*, 102470.

Versfeld, A. (2023). *Making uncertainty: Tuberculosis, substance use, and pathways to health in South Africa.* Rutgers University Press, New Jersey.

Vilone, G., & Longo, L. (2020). Explainable artificial intelligence: a systematic review. *arXiv preprint arXiv:2006.00093.*

Visibelli, A., Roncaglia, B., Spiga, O., & Santucci, A. (2023). The impact of artificial intelligence in the Odyssey of rare diseases. *Biomedicines, 11*(3), 887.

Vrahatis, A. G., Skolariki, K., Krokidis, M. G., Lazaros, K., Exarchos, T. P., & Vlamos, P. (2023). Revolutionizing the early detection of Alzheimer's disease through non-invasive biomarkers: the role of artificial intelligence and deep learning. *Sensors, 23*(9), 4184.

Wang, K., Tian, J., Zheng, C., Yang, H., Ren, J., Liu, Y., ... Zhang, Y. (2021). Interpretable prediction of 3-year all-cause mortality in patients with heart failure caused by coronary heart disease based on machine learning and SHAP. *Computers in Biology and Medicine, 137*, 104813.

Wang, Y., Huang, X., Li, F., Jia, X., Jia, N., Fu, J., ... Huang, S. (2023). Serum-integrated omics reveal the host response landscape for severe pediatric community-acquired pneumonia. *Critical Care, 27*(1), 1–17.

Williams, F., & Jeanetta, S. C. (2016). Lived experiences of breast cancer survivors after diagnosis, treatment and beyond: qualitative study. *Health Expectations, 19*(3), 631–642.

Xu, Z., Yu, F., Zhang, B., & Zhang, Q. (2022). Intelligent diagnosis of left ventricular hypertrophy using transthoracic echocardiography videos. *Computer Methods and Programs in Biomedicine, 226*, 107182.

Yang, J., Soltan, A. A., & Clifton, D. A. (2022). Machine learning generalizability across healthcare settings: insights from multi-site COVID-19 screening. *npj Digital Medicine, 5*(1), 69.

Yang, Q., Suh, J., Chen, N.-C., & Ramos, G. (2018). *Grounding interactive machine learning tool design in how non-experts actually build models.* Paper presented at the Proceedings of the 2018 designing interactive systems conference.

Zameer, R., Tariq, S., Noreen, S., Sadaqat, M., & Azeem, F. (2022). Role of transcriptomics and artificial intelligence approaches for the selection of bioactive compounds. *Drug Design Using Machine Learning*, 283–317.

9 Explainability and the Role of Digital Twins in Personalized Medicine and Healthcare Optimization

İlhan Uysal and Utku Kose

9.1 INTRODUCTION: BACKGROUND AND DRIVING FORCES

Digital twins are essentially digital copies or simulations of real-world physical entities, processes or systems. They mimic the behaviour and dynamics of their real counterparts in real time. These virtual models are created by combining data from various sources such as sensors, Internet of Things (IoT) devices and other data streams into a coherent platform. This platform tries to fulfil the functions of monitoring, analysing and improving the efficiency of the real system it emulates.

The concept of digital twin was first introduced by Dr Michael Grieves in 2002. Grieves defined this concept as a virtual representation of a tangible object or system, which enables the actions and functionality of the object to be predicted. Figure 9.1 provides a graphical representation of the development of the digital twin concept over time. The term gained popularity in the manufacturing industry where it was used to create virtual replicas of industrial equipment to monitor their performance, detect failures and optimize maintenance schedules. Since that time, the use of digital twins has expanded beyond the manufacturing domain and into various industries such as aerospace, automotive and energy. With advances in sensor technology, artificial intelligence (AI), and big data analytics, digital twins have become more sophisticated and capable of simulating complex systems and processes.

The healthcare sector is increasingly recognizing the potential of digital twins in enhancing patient results, streamlining healthcare provision, and cutting down on expenses. By creating virtual models of patients, medical devices, and health processes, digital twins can facilitate personalized medicine, predictive analytics, and real-time monitoring of patient health. For instance, by consolidating data from diverse origins like electronic health records, wearable devices, and genetic examinations, it can generate a digital representation or twin of a patient. This virtual counterpart can be employed to imitate how the patient's body reacts under varying treatments and

DOI: 10.1201/9781003426073-9

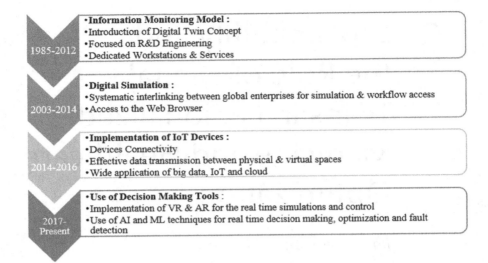

FIGURE 9.1   Evolution of digital twin. (Source: Warke et al., 2021). Infographic showing the evolution of digital twin technology from 1985 to the present. The first period (1985–2002) in red includes key points such as the 'Information Tracking Model', the introduction of the Digital Twin concept, the focus on R&D engineering and dedicated workstations and services. The second period (2003–2014), shown in blue, is 'Digital Simulation', which emphasises the systematic linking of global companies for simulation, workflow access and web browser access. The third period (2014–2016), shown in yellow, is 'Implementation of IoT Devices', which highlights device connectivity, efficient data transmission and the widespread application of big data, IoT and cloud. The last period (2017-present), coloured in green, is 'Use of Decision Making Tools', with the application of AR/VR for real-time simulation and control, and the use of AI and machine learning for real-time decision making, optimisation and fault detection.

Citation in figure caption

actions, forecast the advancement of illnesses, and enhance the strategies for medical care. Digital twins have versatile applications in healthcare, extending to the design and validation of medical equipment, streamlining hospital operations and resource management, and enhancing the precision of clinical decisions. The potential advantages offered by digital twins in the realm of healthcare are vast and have the capacity to revolutionize the provision and administration of healthcare services. Figure 9.2 is a graphical representation showcasing various applications of digital twins, as outlined in the use cases.

To give examples from the studies on digital twins in the literature, in their study, Volkov et al. reviewed existing platforms supporting smart healthcare, in particular digital twins, the IoT, and mobile medicine. They discuss how these technologies can be used to improve healthcare outcomes and present case studies of successful implementations in various healthcare settings (Volkov et al., 2021). Garg

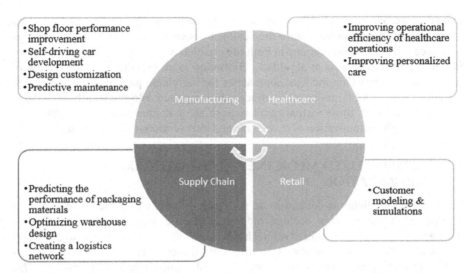

FIGURE 9.2   Uses of the digital twin (Source: Grieves and Vickers, J. 2017). A graphic called 'Use Cases by Industry' for digital twin technology, divided into four sectors. The top left section, 'Manufacturing', focuses on improving shop floor performance, developing driverless vehicles, design customisation and predictive maintenance. Healthcare, top right, focuses on improving operational efficiency and personalised care. Supply Chain, below left, looks at predicting the performance of packaging materials, optimising warehouse design and building a logistics network. The Retail section, below right, highlights customer modelling and simulation. The central cutaway design includes representative images for each sector: robotic arms for manufacturing, a medical stethoscope for healthcare, shipping containers for supply chain and a hand holding a package for retail.

discussed how digital twin technology can revolutionize personalized healthcare. The author explained the benefits of digital twins in healthcare, including personalized medicine and improved patient outcomes, and provided examples of successful implementations (Garg, 2021). Kamel Boulos and Zhang explored the potential uses of digital twins in healthcare, ranging from tailored medical approaches to precise public health strategies. They highlighted the importance of privacy and security when integrating digital twins into healthcare. They also provided examples of successful implementations to illustrate their points (Kamel Boulos and Zhang, 2021). Liu et al. introduced an innovative framework hosted on the cloud that utilizes digital twin technology to enhance elderly healthcare. In this study, they explained how the framework can be used to improve health outcomes for elderly patients and provided an example of successful implementation. Liu et al. (2019) conducted an investigation focused on ethical considerations related to digital twins within the realm of personalized healthcare. Their study meticulously outlined and deliberated upon a range of ethical matters, encompassing aspects such as safeguarding data privacy and security, ensuring informed consent, and addressing the potential for biases (Huang et al., 2022). Bruynseels et al. discussed the ethical implications of using digital twins in healthcare. The research delved into assessing the likely advantages and downsides of employing digital twins within the healthcare sector. It also emphasized the significance of ethical deliberations in both the creation and integration of digital twins in healthcare (Bruynseels et al., 2018). Björnsson et al. explored the application of digital twins in tailoring medical approaches for individual patients. They presented instances where digital twins were effectively integrated into the healthcare system and evaluated how these virtual replicas have the potential to enhance healthcare results through customization of treatments and interventions (Björnsson et al., 2020).

9.2 PERSONALIZED MEDICINE AND HEALTHCARE OPTIMIZATION

Personalized medicine is defined as an approach that aims to personalize treatment approaches by taking into account individual genetic and environmental factors. This approach aims to optimize all healthcare services, from diagnosis to treatment of diseases. Personalized medicine is advancing through the integration of contemporary technologies, extensive data analysis, and sophisticated AI algorithms. This approach is expected to bring benefits such as more accurate diagnosis, more effective treatment, reduced risk of side effects, and more efficient healthcare services. The precision public health cycle, depicted in Figure 9.3, presents a visual representation of how precision methods can contribute to enhancing patient care and overall health outcomes. This cycle highlights the advantages of employing precision approaches in the pursuit of better healthcare for individuals and the wider population (Bilkey et al., 2019).

9.2.1 OVERVIEW OF PERSONALIZED MEDICINE AND ITS BENEFITS

Personalized medicine revolutionizes healthcare by considering an individual's distinct genetic makeup, environmental influences, and lifestyle choices in the diagnosis and treatment of diseases. It recognizes that each patient is inherently unique, necessitating customized approaches to address their specific medical requirements. Customized healthcare, also known as personalized medicine, offers a multitude of advantages, such as precise diagnoses, enhanced treatment efficacy, and improved well-being for patients. Customizing treatments based on individual patient requirements in personalized medicine can mitigate the potential for negative reactions to medications and enhance overall patient contentment (Collins and Varmus, 2015; Tsimberidou and Kurzrock, 2015).

9.2.2 CHALLENGES AND OPPORTUNITIES IN HEALTHCARE OPTIMIZATION

One of the key challenges in healthcare optimization is the integration of different data sources and systems. Healthcare data is often stored in silos and in different formats, making it difficult to integrate and analyze. This issue can be tackled by leveraging digital twin technology, enabling the development of virtual representations that consolidate data from diverse origins. Another challenge is the complexity of healthcare systems and processes. Healthcare systems encompass a variety of participants,

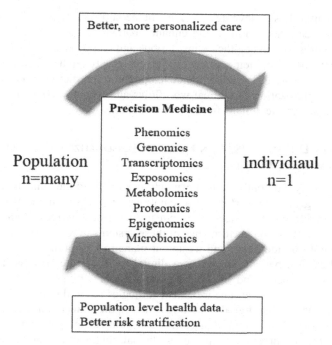

FIGURE 9.3 The cycle of precise public health (Source: Bilkey et al., 2019). Circular diagram illustrating the delicate cycle of public health. The diagram shows a bidirectional flow between 'Population n = many' and 'Individual n = 1' and expresses the dynamic between population-level health data and individual care. The blue arrow moving from the individual to the population is labelled 'Better, more personalised care' and includes the terms 'precision medicine, phenomics, genomics, transcriptomics, exposomics, metabolomics, proteomics, epigenomics and microbiomics'. The pink arrow pointing from the population to the individual is labelled 'Population-level health data'. Better risk stratification. This image summarises the concept of precision public health, where individual data inform broader health strategies and vice versa.

such as patients, healthcare practitioners, insurers, and policymakers, each with their unique aims and purposes. This complexity can make it difficult to identify and implement optimization strategies that benefit all stakeholders. Furthermore, the implementation of healthcare optimization initiatives can be hampered by resistance to change from healthcare providers and patients. Healthcare providers may resist change for fear of disrupting established workflows and practices, while patients may resist changes to care plans or the use of new technologies (Kose and Colakoglu, 2023; Zou et al., 2021; Sun et al., 2023).

Despite these challenges, there are also opportunities for healthcare optimization. An opportunity that stands out is employing AI and machine learning algorithms to examine health-related data, recognizing patterns, and detecting trends. This can facilitate the development of personalized treatment plans and early detection of diseases.

Another potential lies in leveraging telemedicine and remote patient monitoring to enhance healthcare accessibility and enhance patient results. These advancements can alleviate pressure on healthcare institutions and empower patients to access care from the convenience of their residences. Current research in healthcare optimization focuses on several areas, including the development of predictive models for disease progression, optimization of clinical workflows and resource allocation, and determining optimal treatment plans for individual patients.

9.2.3   How Digital Twins Can Facilitate Personalized Medicine and Healthcare Optimization

Digital twins play a pivotal role in advancing personalized medicine and optimizing healthcare by generating virtual representations of patients, medical devices, and healthcare workflows. By integrating data from various sources, digital twins can simulate the behavior and performance of these systems in real time, allowing potential problems to be identified and treatment plans to be optimized. For instance, a digital replica of a patient, known as a digital twin, can be harnessed to forecast how various treatments and interventions might work, pinpoint possible negative responses to drugs, and fine-tune treatment strategies tailored to the individual needs of the patient. Moreover, digital twins find utility in the conceptualization and experimentation of medical apparatus and gear, streamlining hospital operations and effectively allocating resources, and enhancing clinical judgment. Through the emulation of the actions and efficacy of these systems, digital twins can effectively detect potential issues and enhance the efficiency of healthcare provision.

9.3   APPLICATIONS OF DIGITAL TWINS IN HEALTHCARE

Digital twins involve creating virtual representations of real-world entities and are valuable for monitoring, predicting, and enhancing the functionality of these entities. They essentially constitute virtual replicas of physical objects or systems. In the medical domain, digital twins can be harnessed to construct personalized models of an individual's anatomy. The integration of digital twins within healthcare holds immense potential, influencing various aspects such as patient treatment, surgical procedures, efficient operation and oversight of medical equipment, patient data management, disease prognosis, and advancements in medical research. Envisaged applications of digital twins encompass elevating patient outcomes, enriching medical education, and mitigating healthcare expenditures (Sun et al., 2022; Gaebel et al., 2021; Angulo et al., 2020). Figure 9.4 illustrates the diverse applications of digital twins within the healthcare sector.

9.3.1   Disease Prediction and Early Diagnosis

One of the primary uses of digital twins in the healthcare sector involves predicting diseases and detecting them at an early stage. Digital twins achieve this by examining a variety of data sources, including electronic health records, medical imaging, and data from wearable devices. By doing so, they can recognize patterns and trends

Disease Prediction and Early Diagnosis

Treatment Optimization and Individualized Theraphy

Medical Device and Equipment Design and Testing

Healthcare Process and Workflow Optimization

FIGURE 9.4 Applications of Digital Twins in Healthcare. Collage of four applications of digital twins in healthcare, each represented by a circular image. The top left shows 'Disease Prediction and Early Diagnosis' with an image of a brain and a neural network, symbolising artificial intelligence in diagnostics. Top right, 'Treatment Optimisation and Individualised Therapy' with two doctors observing a digital twin on the screen. The bottom left is 'Medical Device and Equipment Design and Testing', which focuses on advanced machines and virtual testing. On the bottom right is 'Healthcare Process and Workflow Optimisation', where healthcare professionals participate in a coordinated environment with digital displays. Each image conveys the integration of technology and healthcare to improve patient outcomes.

that might signify the presence of a disease or the potential risk of developing one. To illustrate, a patient's digital twin could be utilized to analyze their medical background, lifestyle choices, and genetic data, enabling predictions regarding the likelihood of the patient developing specific conditions such as heart disease or cancer. Detecting these conditions early on can result in more efficient treatments and improved outcomes for the patients.

In 2021, Allen et al. introduced an advanced machine learning model using digital twin technology to forecast the progression of strokes in individuals. They stated that the model can accurately predict a patient's likelihood of developing stroke-related complications based on various clinical and demographic data (Allen et al., 2021). In their research on the utilization of digital twin technology for anticipating neurological issues and therapies in pediatric cancer cases, Thiong'o et al. highlighted the capability of digital twins in generating individualized brain models for patients. These models play a crucial role in forecasting potential complications and providing valuable insights to guide treatment choices (Thiong'o and Rutka, 2022). In another 2021 study, Eleyan et al. introduced a digital twin strategy to develop intelligent healthcare systems with contextual awareness using IoT devices. The authors claimed that their suggested method enables ongoing patient monitoring in real time, ultimately enhancing the effectiveness of healthcare services (Eleyan et al., 2021). In a study conducted in 2020, Bertolini et al. introduced a digital twin methodology to model the advancement of diseases in individuals experiencing mild cognitive impairment and Alzheimer's disease. They stated that their approach could help clinicians make more informed decisions about patient care and improve outcomes. Bertolini et al. (2020). introduced a digital twin methodology for constructing clinical decision support systems (DSS) utilizing explainable AI. In this study, they stated that their approach can improve the accuracy and transparency of DSS and provide more personalized patient care (Rao and Mane, 2019).

9.3.2 TREATMENT OPTIMIZATION AND INDIVIDUALIZED THERAPY

Another way digital twins are being used in healthcare is to optimize treatments and tailor therapy to individual patients. By creating a virtual representation of a patient, digital twins can mimic the effects of various treatments and interventions, enabling healthcare professionals to determine the most effective approach for each patient. For example, in the case of a cancer patient, the digital twin can simulate the results of different chemotherapy regimens and help healthcare providers develop the most effective treatment strategy tailored to the patient's different characteristics.

The utilization of digital twins is believed to enhance the efficiency and efficacy of clinical trials, in addition to aiding in the creation of personalized treatment strategies for individual patients (Sinisi et al., 2020). Digital twins have the potential to observe and assess patient actions, enhance results, and offer tailored interventions (Lauer-Schmaltz et al., 2022). Digital twins have the potential to tailor treatment strategies for individuals dealing with persistent conditions such as diabetes or hypertension. Through analyzing data from wearable devices and various sources, digital twins allow experts to modify treatment strategies as needed in relation to a patient's health condition.

9.3.3 MEDICAL DEVICE AND EQUIPMENT DESIGN AND TESTING

Digital twins also find application in the design and testing of medical equipment and devices. By creating virtual representations of these devices, digital twins can mimic their movement and function in different situations. This simulation capability can enable manufacturers to identify potential challenges and improve their designs before the production phase. For example, a digital twin of a new medical device could simulate how it works with various patient demographics, allowing manufacturers to identify potential problems and fine-tune its efficiency before it is released to the market.

9.3.4 HEALTHCARE PROCESS AND WORKFLOW OPTIMIZATION

Digital twins have the capability to enhance the efficiency of healthcare processes and workflows. Through simulation of how these systems function and perform, digital twins can pinpoint possible bottlenecks and ineffective areas. This empowers healthcare providers to streamline their operations and enhance patient results. For instance, creating a digital twin for a hospital allows for the simulation of patient, staff, and equipment movement within the facility. This simulation helps healthcare providers foresee issues and optimize their operations to boost efficiency and enhance patient contentment.

9.4 CASE STUDIES

9.4.1 Examples of Successful Implementation of Digital Twins in Healthcare

Virtual Heart: A team of researchers at the University of Sheffield has created a digital replica of the heart that can be used to simulate different treatments and interventions to understand their impact on people with heart disease (Highfield, 2023). The virtual heart is based on data from medical imaging and is personalized to each patient's unique anatomy and physiology, allowing healthcare providers to develop individualized treatment plans.

The Brain's Digital Twin: Scientists are actively working to create a digital replica of the brain to simulate the effects of different treatments and interventions for people with neurological conditions such as epilepsy and Parkinson's disease. The digital twin is personalized to each patient's unique brain structure and function based on data from medical imaging, enabling healthcare providers to develop individualized treatment plans (Sinisi et al., 2020).

Virtual Intensive Care Unit: A team of researchers from Johns Hopkins University has created a digital replica of an intensive care unit (ICU) to improve patient care and outcomes. This virtual recreation of an ICU uses information from electronic health records to mimic the actions and efficiency of a real ICU, allowing medical professionals to pinpoint potential problems and improve overall operational efficiency (Topol, 2019).

9.4.2 Real-World Use Cases of Personalized Medicine and Healthcare Optimization Using Digital Twins

The practical applications of personalized medicine and healthcare optimization using digital twins are broad and full of potential. These digital counterparts, derived from data from various medical streams such as imaging, genetic profiling, electronic health records, and more have the ability to replicate and simulate the functions and efficiency of both the human body and healthcare facilities in real time. The use of digital twins could pave the way for more precise treatment strategies, improved patient outcomes, and streamlined healthcare processes.

Personalized Cancer Therapy: Researchers are actively working to create a digital replica of a patient's tumour, enabling the simulation of different cancer treatments and the identification of the optimal approach for individual patients.

This digital twin is created using data from medical imaging and genetic testing, allowing healthcare professionals to tailor cancer treatment strategies to each patient's individual needs (Dumbrava and Meric-Bernstam, 2018).

Optimizing Hospital Operations: A digital replica of a hospital has been created by researchers at the Mayo Clinic. This innovative tool is designed to improve hospital operations and patient outcomes. Using data from electronic health records, the digital twin can accurately model the actions and effectiveness of the hospital in real time. This capability enables healthcare professionals to identify potential barriers and inefficiencies, allowing them to streamline operations and improve overall performance (Liu et al., 2019).

Personalized Diabetes Management: Diabetes is a chronic disease that requires continuous management to prevent complications. Personalized diabetes management has emerged as a promising approach to optimize diabetes care. One of the recent technological advances that have the potential to enhance personalized diabetes management is the use of digital twin technology. Digital twin technology involves creating a digital copy of a real-world entity, such as a person with diabetes, to simulate different situations and predict potential outcomes. In their paper, Jensen and colleagues presented Diasnet Mobile, a customized mobile platform for diabetes monitoring and advice. This platform uses a digital twin to provide immediate guidance to people with diabetes. The digital twin is built using data from glucose monitoring devices, insulin pumps and several other data sources. The system was evaluated in a pilot study and showed promising results in improving glycemic control (Jensen et al., 2007). Subramanian and Hirsch discussed the need for individualized diabetes treatment and proposed the use of digital twin technology to achieve this goal. They emphasized the importance of integrating data from multiple sources such as medical records, continuous glucose monitoring devices, and activity trackers to develop a comprehensive digital twin (Subramanian and Hirsch, 2014). Williams et al. provided an update on personalized approaches to type 2 diabetes management, while also exploring the potential of digital twin technology to improve diabetes care. They stated the importance of incorporating patient preferences and values into the digital twin to ensure patient-centered care (Williams et al., 2022). O'Hair highlighted the future impact of digital twin technology in tailoring diabetes care, emphasizing the critical integration of behavioral and environmental data into the digital twin for a more comprehensive approach (O'Hair, 2013). Fico et al. conducted a preliminary investigation into incorporating personalized healthcare pathways into an information and communication technology (ICT) system designed for diabetes management. Within this platform, a digital twin was used to replicate different situations and predict potential outcomes. The results of the study showed that digital twin technology holds promise for improving diabetes care (Fico et al., 2014). Park et al. developed a personal diabetes management system using a digital twin and designed for point-of-care use on devices. This system provides personalized advice to people with diabetes. The digital twin is based on data from glucose monitoring devices, insulin pumps, and other sources. The system was evaluated in a pilot study and showed promising results in improving glycemic control (Park et al., 2006). Lee et al. introduced a personalized approach focused on treatment outcomes for diabetes management. They used a digital twin to simulate different treatment strategies and predict potential outcomes. The researchers stated that it is important to incorporate patient preferences and values into the digital twin to ensure patient-centered care (Lee et al., 2018).

9.5 CHALLENGES AND FUTURE DIRECTIONS

9.5.1 Technical and Ethical Challenges in Implementing Digital Twins in Healthcare

Data Privacy and Security: A major hurdle in the integration of digital twins into healthcare revolves around the privacy and security of patient data. Digital twins require large amounts of sensitive patient data, including medical images, genetic information, and electronic health records. As a result, healthcare providers must ensure that this data is stored securely and that only authorized individuals have access to it. Popa et al. highlighted key socio-ethical benefits associated with the use of digital twins in healthcare, including increased patient empowerment, personalized care, and, ultimately, improved health outcomes. However, they also recognized potential socio-ethical risks, such as data privacy breaches and discrimination (Popa et al., 2021).

Data Integration and Interoperability: Another challenge in implementing digital twins in healthcare is integrating data from different sources and ensuring interoperability. Digital twins require the aggregation of data from disparate sources such as medical imaging, electronic health records, wearable devices, and more. As a result, healthcare organizations need to validate data compatibility and streamline its integration into a unified platform. Lutze highlighted the importance of standardization and data governance in ensuring data integration and interoperability. Establishing protocols for sharing data and implementing effective data management practices can streamline the amalgamation of information from various origins, allowing for the construction of holistic patient profiles (Lutze, 2019).

Technical Complexity and Cost: Developing and implementing digital twins in healthcare requires advanced technical expertise and significant financial investment. The development and maintenance of digital twins require specialized software, hardware, and personnel, making it challenging for smaller healthcare organizations to implement. Armeni et al. highlighted the importance of conducting a cost-benefit analysis when creating and integrating digital twins. In the healthcare setting, it is important for healthcare providers to carefully weigh the benefits of digital twins against the costs associated with their implementation. These costs include hardware, software, and staff training and support (Armeni et al., 2022).

Ethical Considerations: The implementation of digital twins in the healthcare sector raises ethical considerations surrounding patient consent, data ownership, and potential biases in algorithms. It is crucial for healthcare providers to guarantee that patients are well-informed about the utilization of their data and have provided informed consent. Moreover, it is essential to ensure that the algorithms integrated into digital twins are fair and do not perpetuate existing healthcare inequalities. In their work, Drummond and Coulet underscored the significance of obtaining informed consent during the development and deployment of digital twins. Healthcare providers must make certain that patients have a comprehensive understanding of how their data is gathered and utilized, and they should seek explicit consent from patients before utilizing their data (Drummond and Coulet, 2022).

9.5.2 FUTURE DIRECTIONS FOR RESEARCH AND DEVELOPMENT IN DIGITAL TWINS FOR HEALTHCARE

Improved Data Integration and Interoperability: Future advances in the use of digital twins in healthcare should priorities improving the seamless integration and interoperability of data. Healthcare providers and stakeholders should work to establish standardized data formats and protocols to facilitate the integration of data from diverse sources into a unified platform.

Advanced Machine Learning and Artificial Intelligence: Machine learning and AI algorithms of a sophisticated nature are assuming a progressively significant function in the exploration and enhancement of digital twins tailored for healthcare purposes in the forthcoming era. These advanced algorithms make it possible to examine large datasets, enabling healthcare providers to recognize complex patterns and correlations. Using this information, healthcare providers can develop highly accurate and personalized treatment plans for patients. Digital twins, which are virtual replicas of individuals or systems, are created by merging patient data, including electronic health records, medical imaging, and genetic details. Machine learning algorithms powered by AI are used to analyze these complex datasets and derive meaningful insights. These algorithms can efficiently identify subtle correlations, predict patient outcomes, and recommend optimal interventions. The use of advanced machine learning and AI techniques in digital twins offers unprecedented opportunities for healthcare. Through ongoing real-time data monitoring and updates to the digital twin, healthcare providers can simulate various treatment approaches and forecast their impacts. This proactive method enables the early identification of potential health issues and the customization of treatments for each patient, ultimately enhancing the quality of care (Kaur et al., 2020). Here, it may be also important to apply intelligent optimization solutions since they have real-time solution capabilities for better management and enhanced data (Guraksin et al., 2019, Yigit et al., 2018). In addition to treatment planning, advanced machine learning algorithms can contribute to disease diagnosis and prognosis. By comparing data from an individual's digital twin with large-scale population datasets, machine learning models can detect early signs of diseases, identify risk factors and even predict the likelihood of future medical conditions. Such insights empower healthcare providers to initiate preventive measures and intervene at the earliest stages of disease progression. Thus, the integration of advanced machine learning and AI algorithms into digital twins for healthcare holds great promise. The potential for this technology to transform the sector is immense, providing tailored insights based on data and equipping healthcare professionals with valuable resources to make informed decisions. This in turn improves patient outcomes and drives advances in medical research.

Explainability: In AI applications, digital twins help us understand how complex AI models make decisions and their consequences. Explainability provides an understanding of the internal working mechanisms and decision-making processes of these AI models, allowing us to assess the causes and effects of their decisions. Digital twins help to understand complex AI systems by simulating and visualizing AI models, explaining the basis of their decisions, what data is influential and how this data is reflected in the results. This understanding contributes to making algorithms and models more transparent, reliable and ethically acceptable. Especially in healthcare, the ability of digital twins to explain the work of AI models is of great importance in disease diagnosis, treatment recommendations, and interpretation of health outcomes. In this way, healthcare professionals can better understand the recommendations and decisions of these models and use this information to make clinical decisions. At the same time, it will enable patients and other stakeholders to trust in this process and encourage a more transparent approach to healthcare.

Application of Blockchain Technology: The use of blockchain technology can ensure the confidentiality and security of patient information within digital twin systems. By applying blockchain technology, healthcare providers can securely store patient data and restrict access to authorized individuals.

Increased Collaboration and Standardization: Collaboration between healthcare providers and technology developers is essential to establish unified standards and guidelines for crafting and integrating digital twins within the healthcare sector. This joint effort and standardization are pivotal in enhancing the operational efficiency and overall effectiveness of utilizing digital twins in healthcare.

9.6 CONCLUSION

Digital twins hold great promise for transforming personalized medicine and improving healthcare efficiency. They can revolutionize early disease detection, tailor treatments to individuals, optimize medical device design, and streamline healthcare processes. Through the use of advanced machine learning and AI algorithms, healthcare providers can develop highly accurate personalized treatment strategies, ultimately leading to better patient outcomes and reduced healthcare costs.

The anticipated impact of digital twins on the healthcare landscape is profound. By harnessing the power of digital twins, healthcare providers can offer more accurate diagnoses, personalized treatment plans, and optimized healthcare procedures, resulting in improved patient outcomes and cost savings. In addition, digital twins have the potential to accelerate the development of novel medical devices and treatments, ultimately improving patient care and outcomes. However, the integration of digital twins into healthcare is fraught with challenges, including data security and privacy, seamless data integration, technical complexity, cost, and ethical concerns. Collaboration between healthcare providers and technology developers is essential to overcome these barriers and ensure the responsible development and integration of digital twins in healthcare.

In conclusion, the importance of digital twins, especially their accountability, in personalized medicine and healthcare optimization cannot be overstated. With continued research and development, digital twins will improve patient outcomes, reduce healthcare costs, and ultimately save lives.

REFERENCES

Allen, A., Siefkas, A., Pellegrini, E., Burdick, H., Barnes, G., Calvert, J., Das, R. (2021). A digital twins machine learning model for forecasting disease progression in stroke patients. *Applied Sciences*, 11(12), 5576.

Angulo, C., Gonzalez-Abril, L., Raya, C., Ortega, J. A. (2020, April). A proposal to evolving towards digital twins in healthcare. In *Bioinformatics and Biomedical Engineering: 8th International Work-Conference, IWBBIO 2020, Granada, Spain, May 6–8, 2020, Proceedings* (pp. 418–426). Cham: Springer International Publishing.

Armeni, P., Polat, I., De Rossi, L. M., Diaferia, L., Meregalli, S., Gatti, A. (2022). Digital twins in healthcare: is it the beginning of a new era of evidence-based medicine? A critical review. *Journal of Personalized Medicine*, 12(8), 1255.

Bertolini, D., Loukianov, A. D., Smith, A. M., Li-Bland, D., Pouliot, Y., Walsh, J. R., Fisher, C. K. (2020). Modeling disease progression in mild cognitive impairment and Alzheimer's disease with digital twins. *arXiv e-prints* doi: 10.48550/arXiv.2012.13455.

Bilkey, G. A., Burns, B. L., Coles, E. P., Mahede, T., Baynam, G., Nowak, K. J. (2019). Optimizing precision medicine for public health. *Frontiers in Public Health*, 7, 42.

Björnsson, B., Borrebaeck, C., Elander, N., Gasslander, T., Gawel, D. R., Gustafsson, M., Swedish Digital Twin Consortium. (2020). Digital twins to personalize medicine. *Genome Medicine*, 12, 1–4.

Bruynseels, K., Santoni de Sio, F., Van den Hoven, J. (2018). Digital twins in health care: ethical implications of an emerging engineering paradigm. *Frontiers in Genetics*, 9, 31.

Collins, F. S., Varmus, H. (2015). A new initiative on precision medicine. *New England Journal of Medicine*, 372(9), 793–795.

Drummond, D., Coulet, A. (2022). Technical, ethical, legal, and societal challenges with digital twin systems for the management of chronic diseases in children and young people. *Journal of Medical Internet Research*, 24(10), e39698.

Dumbrava, E. I., Meric-Bernstam, F. (2018). Personalized cancer therapy-leveraging a knowledge base for clinical decision-making. *Cold Spring Harbor molecular Case Studies*, 4(2), a001578. https://doi.org/10.1101/mcs.a001578.

Elayan, H., Aloqaily, M., Guizani, M. (2021). Digital twin for intelligent context-aware IoT healthcare systems. *IEEE Internet of Things Journal*, 8(23), 16749–16757.

Fico, G., Fioravanti, A., Arredondo, M. T., Gorman, J., Diazzi, C., Arcuri, G., Pirini, G. (2014). Integration of personalized healthcare pathways in an ICT platform for diabetes managements: a small-scale exploratory study. *IEEE Journal of Biomedical and Health Informatics*, 20(1), 29–38.

Gaebel, J., Keller, J., Schneider, D., Lindenmeyer, A., Neumuth, T., Franke, S. (2021). The digital twin: modular model-based approach to personalized medicine. *Current Directions in Biomedical Engineering*, 7(2), 223–226.

Garg, H. (2021). Digital twin technology: Revolutionary to improve personalized healthcare. *Science Progress and Research (SPR)*, 1(1), 32–34. https://doi. org/10.52152/spr/2021.105.

Grieves, M., Vickers, J. (2017). Digital twin: Mitigating unpredictable, undesirable emergent behavior in complex systems, in: Transdisciplinary Perspectives on Complex Systems. Springer, pp. 85–113.

Guraksin, G.E., Deperlioglu, O., Kose, U. (2019). A Novel Underwater Image Enhancement Approach with Wavelet Transform Supported by Differential Evolution Algorithm. In: Hemanth, J., Balas, V. (eds) Nature Inspired Optimization Techniques for Image Processing Applications. Intelligent Systems Reference Library, vol 150. Springer, Cham. https://doi.org/10.1007/978-3-319-96002-9_11.

Highfield, R. (2023). Virtual You : How Building Your Digital Twin Will Revolutionize Medicine and Change Your Life. Princeton University Press, New Jersey. Retrieved from http://digital.casalini.it/9780691223407.

Huang, P. H., Kim, K. H., Schermer, M. (2022). Ethical issues of digital twins for personalized health care service: preliminary mapping study. *Journal of Medical Internet Research*, 24(1), e33081.

Jensen, K. L., Pedersen, C. F., & Larsen, L. B. (2007). Diasnet Mobile - A Personalized Mobile Diabetes Management and Advisory Service. Paper presented at Second International Workshop on Personalisation for e-Health in conjunction with User Modelling 2007, Corfu, Greece. www.iit.demokritos.gr/um2007/UM2007_WS7_PEL.pdf.

Kamel Boulos, M. N., Zhang, P. (2021). Digital twins: from personalized medicine to precision public health. *Journal of Personalized Medicine*, 11(8), 745.

Kaur, M.J., Mishra, V.P., Maheshwari, P. (2020). The Convergence of Digital Twin, IoT, and Machine Learning: Transforming Data into Action. In: Farsi, M., Daneshkhah, A., Hosseinian-Far, A., Jahankhani, H. (eds) Digital Twin Technologies and Smart Cities. Internet of Things. Springer, Cham. https://doi.org/10.1007/978-3-030-18732-3_1

Kose, G., Colakoglu. O. E. (2023). Health tourism with data mining: Present state and future potentials. *International Journal of Information Communication Technology and Digital Convergence*, 8(1), 23–33.

Lauer-Schmaltz, M. W., Cash, P., Hansen, J. P., Maier, A. (2022). Designing human digital twins for behaviour-changing therapy and rehabilitation: a systematic review. *Proceedings of the Design Society*, 2, 1303–1312.

Lee, E. K., Wei, X., Baker-Witt, F., Wright, M. D., Quarshie, A. (2018). Outcome-driven personalized treatment design for managing diabetes. *Interfaces*, 48(5), 422–435.

Liu, Y., Zhang, L., Yang, Y., Zhou, L., Ren, L., Wang, F., Deen, M. J. (2019). A novel cloud-based framework for the elderly healthcare services using digital twin. *IEEE Access*, 7, 49088–49101.

Lutze, R. (2019).Digital Twins in eHealth: Prospects and Challenges Focussing on Information Management. 2019 IEEE International Conference on Engineering, Technology and Innovation (ICE/ITMC), Valbonne Sophia-Antipolis, France, 2019, pp. 1–9, doi: 10.1109/ICE.2019.8792622.

O'Hair A.K., Personalized Diabetes Management, Jun. 2013, Massachusetts Institute of Technology, Doctoral Thesis in, 111 pages.

Park, K. S., Kim, N. J., Hong, J. H., Park, M. S., Cha, E. J., Lee, T. S. (2005). PDA based point-of-care personal diabetes management system. 2005 IEEE Engineering in Medicine and Biology 27th Annual Conference, Shanghai, China, 2005, pp. 3749–3752, doi: 10.1109/IEMBS.2005.1617299.

Popa, E. O., van Hilten, M., Oosterkamp, E., Bogaardt, M. J. (2021). The use of digital twins in healthcare: socio-ethical benefits and socio-ethical risks. *Life Sciences, Society and Policy*, 17(1), 1–25.

Rao, D., Mane, S. (2019). Digital Twin Approach to Clinical DSS With Explainable Ai. Available at: https://arxiv.org/abs/1910.13520v1.

Sinisi, S., Alimguzhin, V., Mancini, T., Tronci, E., Mari, F., Leeners, B. (2020). Optimal personalised treatment computation through in silico clinical trials on patient digital twins. *Fundamenta Informaticae*, 174(3–4), 283–310.

Subramanian, S., Hirsch, I. B. (2014). Personalized diabetes management: moving from algorithmic to individualized therapy. *Diabetes Spectrum*, 27(2), 87–91.

Sun, T., He, X., Li, Z. (2023). Digital twin in healthcare: Recent updates and challenges. *Digital Health*, 9, 20552076221149651.

Sun, T., He, X., Song, X., Shu, L., Li, Z. (2022). The digital twin in medicine: A key to the future of healthcare? *Frontiers in Medicine* (Lausanne). 2022 Jul 14;9 :907066. doi: 10.3389/fmed.2022.907066.

Thiong'o, G. M., Rutka, J. T. (2022). Digital twin technology: the future of predicting neurological complications of pediatric cancers and their treatment. *Frontiers in Oncology*, 11, 781499.

Topol, E. J. (2019). High-performance medicine: the convergence of human and artificial intelligence. *Nature Medicine*, 25(1), 44–56.

Tsimberidou, A. M., Kurzrock, R. (2015). Precision medicine: lessons learned from the SHIVA trial. The Lancet Oncology, 16(16), e579–e580.

Volkov, I., Radchenko, G., Tchernykh, A. (2021). Digital twins, internet of things and mobile medicine: a review of current platforms to support smart healthcare. *Programming and Computer Software*, 47, 578–590.

Warke, V., Kumar, S., Bongale, A., Kotecha, K. (2021). Sustainable development of smart manufacturing driven by the digital twin framework: A statistical analysis. *Sustainability*, 13(18), 10139.

Williams, D. M., Jones, H., Stephens, J. W. (2022). Personalized type 2 diabetes management: an update on recent advances and recommendations, diabetes, metabolic syndrome and obesity. *Diabetes, Metabolic Syndrome and Obesity: Targets and Therapy*, 15, 281–295, DOI: 10.2147/DMSO.S331654.

Yigit, T., Unsal, O., Deperlioglu, O. (2018). Using the metaheuristic methods for real-time optimisation of dynamic school bus routing problem and an application. *International Journal of Bio-Inspired Computation*, 11(2), 123–133.

Zou, P., Shao, J., Luo, Y., Thayaparan, A., Zhang, H., Alam, A., Sidani, S. (2021). Facilitators and barriers to healthy midlife transition among South Asian immigrant women in Canada: A qualitative exploration. *Healthcare*, 9(2):182. https://doi.org/10.3390/healthcare9020182.

10 XAI for Trustworthiness in Medical Tourism

Gamze Kose and Utku Kose

10.1 INTRODUCTION

Today's world includes many advanced forms of smart systems that run integrated data analyzing and processing techniques using artificial intelligence (AI). Artificial intelligence is a field that hosts advanced forms of algorithms to achieve adaptive problem solutions and inferencing mechanisms to learn from the problem data (Hopgood, 2021; Joshi, 2020; Michalski et al., 2013). It is critical that such algorithmic infrastructure of AI is combined with the necessary communication steps for advancing the outcomes accordingly. Artificial intelligence-based smart systems are used in all fields of modern life for automated recommendations, future predictions, designing new data, and for accurate planning (Gomes, 2000; Jackson, 2019; Jiang et al., 2021; Michalski et al., 2013; Pannu, 2015). In detail, these problems are shaped according to input-output structure of specific fields so that smart systems have the role of transforming input data to the outputs that can be used on the user side. However, as a result changing conditions of life real-world problems have been changing dramatically. New problem types have been appearing as a result of advanced technology use and data production. As a result, AI-based systems also become complicated.

In order to ensure successful results for even the most challenging problems, advanced design of AI algorithms and development of stronger architectures have been important for the user side. However, complex features and mechanisms of smart systems have an increased number of parameters to learn and components to analyze to create connections among input and output data. Unfortunately, a human (user) cannot track and determine if the number of parameters is very high. Thus, the field of AI aims at the interpretability of artificial intelligence, a new field called explainable artificial intelligence (XAI). XAI is actually the extended version of interpretable AI and ensures new methodologies enabling the AI model to explain to the user how it processed the data (Arrieta et al., 2020; Confalonieri et al., 2021). In this way, it becomes easy to understand if the model has some inside errors, or requires maintenance in terms of data or model features. Therefore, XAI allows increases in trust level for further applications (Druce et al., 2021; Guo, 2020).

Use of XAI is an effective way to build trustworthy smart systems. This is critical especially in vital fields where the human factor and the data have importance for successful outcomes. For example, XAI has been widely used for healthcare and medical problems, in order to build trustworthy smart systems that can explain how they perform diagnosis, detect abnormal biological factors, and aid medical staff decision making (Loh et al., 2022; Mohanty & Mishra, 2022). Similarly, XAI is effective in other fields where careful data analysis is important for desired outcomes (Arrieta et al., 2020; Khosravi et al., 2022; Mehdiyev & Fettke, 2021). In AI applications, human factor has an effective role in determining the way of problem solution. As a result of using different mobile technologies and platforms to obtain human data, the number of integrated technologies, and AI-based solutions have increased in last decade. Thus, it is important to understand how XAI can affect trust building in different kinds of problems related to human data use. By providing both healthcare and tourism experiences to health tourists, the health tourism area has the potential in artificial intelligence-based applications. In this context, it has a valuable place in also XAI.

The objective of this chapter is to provide an overview of how XAI can be used in health tourism applications. Specifically, we discuss how knowledge of XAI for technical healthcare and medical applications can be transferred to health tourism areas. This chapter aims at enabling interested researchers to understand how XAI can be used in medical tourism problems, by considering different user types from health tourists to health staff or service providers.

The chapter is organized as follows: The next section provides general explanations regarding the concepts of health tourism as well as medical tourism. Additionally, it provides some explanations about how AI can be used for the purposes of medical tourism. In this way, the section establishes the connection with the third section where XAI use for medical tourism is explained. In the third section, application scenarios for XAI in medical tourism are discussed in detail to understand not only technical aspects but also different components in a typical application. Following to this section, the fourth section includes discussion about present and future states, and finally, the chapter ends by providing conclusions in the last section.

10.2   HEALTH AND MEDICAL TOURISM

Health tourism is associated with the travels made to another country for receiving healthcare (Amouzagar, 2016; Reisman, 2010; Smith & Puczkó, 2014). As a result of combining healthcare needs and tourism-related activities, health tourism has become a popular activity worldwide. Since health tourism includes patients (health tourists), medical staff, and service providers, applications should address all these users. Health tourists (Majeed et al., 2018; Sag & Sengul, 2019) engage in health tourism activities including treatments and tourism-oriented trips. Health tourism has the following advantages (Aydin & Karamehmet, 2017; Kilavuz, 2018; Reisman, 2010):

- Health tourists can receive medical operations as well as treatments and take part in well-being activities.
- Health tourists visit different countries not only for healthcare purposes but also for different cultural travel experiences.

- Health tourists can receive better healthcare in countries with better conditions.
- Health tourists can receive healthcare with lower costs when compared with their home countries.
- Medical staff can have experience in terms of different kinds of treatment needs from international patients.
- Service providers (health tourism agencies or tourism agencies) can increase their incomes as a result of receiving health tourists from countries with better financial conditions.

Typical health tourism activities are performed by health tourists, medical staff, and service providers. Health tourists receive healthcare and participate in tourism activities. They are the target audience of health tourism. Medical staff generally includes doctors and the supportive staff that work in hospitals or clinics to perform healthcare tasks for health tourists. On the other hand, service providers include the staff working specifically to plan and sell health tourism programs or re-design ongoing tourism activities according to the scope of health tourism. Since health tourism includes interactions among all these actors, it is highly affected by advancements in terms of information and communication technologies.

Health tourism can be divided into several sub-areas such as disabled tourism, elderly tourism, thermal tourism, spa-wellness tourism, and medical tourism (Isa et al., 2020; Ormond, 2014; Smith & Puczkó, 2014). Among these sub-areas, medical tourism provides medical services (e.g., operations, health treatments, etc.) to health tourists.

10.2.1 MEDICAL TOURISM

Medical tourism includes travel for medical care such as operations (surgeries), dental or cosmetic treatments (Connell, 2006; Moghavvemi et al., 2023; Nisa & Sharma, 2023), and may include special medical equipment, medical devices, and software systems (e.g., for screening, diagnosing, or testing) to be used by staff. This means medical tourism applications may require building special clinical environments and training staff to work with them (Figure 10.1). Moreover, close interaction with technological tools make AI an active component for medical tourism applications.

While medical tourism is considered as a general ecosystem, data processing and communication aspects are run differently for health tourists, service providers, and medical staff. However, the general ecosystem should include data synergy for the different users involved. Figure 10.2 shows a sample technological interaction flow that may be applied for medical tourism applications.

As can be seen in Figure 10.2, the modular structure may be complicated due to different technological components running collaboratively. Furthermore, analysis of user data requires smart systems based on AI algorithms. These smart systems may be simpler to interpret when one or two AI algorithms are running in separate roles. But multiple algorithms running in a complex flow may make the system difficult to be interpreted by staff tracking AI-supported medical tourism events. Additionally, today's advanced AI models (e.g., deep learning) have many parameters to learn,

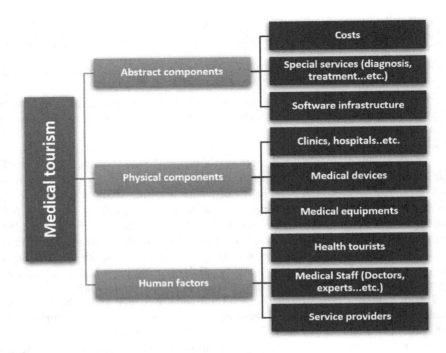

FIGURE 10.1 Components shaping medical tourism.

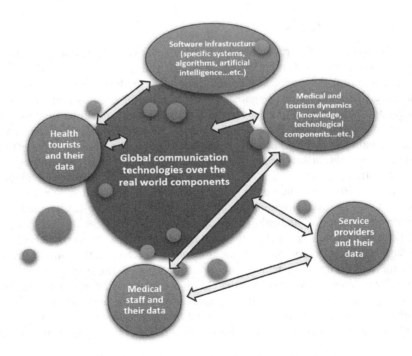

FIGURE 10.2 A sample technological interaction flow for medical tourism.

which make them black box for human use. Because of this, they should be supported with XAI methods to improve users' understanding and increase trust in smart systems.

10.3　XAI FOR ENSURING TRUST IN MEDICAL TOURISM APPLICATIONS

XAI provides different mechanisms to track model parameters, which are pointing how input data is processed to obtain the output data (in other terms, the decision made by the model). In this way, the following advantages are ensured for medical tourism applications:

- Black-box model mechanisms are made transparent to the user side. So, automated decisions made for health tourists, medical staff or service providers can be checked for any algorithmic error. For health tourists, software applications can be provided with transparent explanations so they can feel safe while using such algorithm-based tools.
- Specific data analysis and processing steps inside a complex model can be tracked easily from the user side. So, service providers can make it sure their AI-based software systems are running properly or requiring error corrections by developers. Similarly, doctors and medical staff can understand if AI-based systems are running correctly while processing medical data.
- The model is able to give feedback regarding how it achieves the connections between input and output data. In this way, medical staff and service providers can track decision mechanisms of AI algorithms and step in anytime if there is a problem during data processing steps.
- For the user side, it becomes easy to track errors and sustainability of the model. For medical tourism applications, this advantage allows less software related maintenance so the focus will be forwarded better to activities.
- Users can receive early alerts against problems inside the running model. So, active medical tourism events can run effectively while technical precautions for AI infrastructure can be taken earlier on the background.
- From a general perspective, a safe model improves user's trust level. This is important for all actors of a medical tourism to run activities without having any anxieties against used software systems.
- Thanks to XAI, it is possible to create safe models for developing human-compatible smart systems. In this way, medical tourism activities can be better automated through AI-based systems.

As it can be understood from the listed advantages, a wide variety of medical tourism applications can be supported with XAI components to achieve the desired trustworthiness. Such use of XAI will ensure model sustainability too.

When a desired level of trust is ensured in a medical tourism application, the following advantages are seen:

- Service providers know better about how smart systems can be used to target health tourists and plan specific activities.

- Data belonging to health tourists can be used effectively without any bias or errors inside the smart systems.
- It becomes easier to use digital tools (of the built software system), even when the medical tourism application employs the most advanced smart systems.
- For global ecosystems of medical tourism, it becomes safe to track and manage all smart systems from one point of view.
- Health tourists can feel comfortable while receiving instant, transparent reports about smart systems.
- Medical staff has a desired trust level for smart systems and the service providers of the medical tourism application they use.
- Digital components of the whole medical tourism ecosystem are protected against data- and AI-related problems.

10.3.1 XAI Methodologies with Potential Uses

There are different kinds of XAI methods that can be used in medical tourism applications to expand interpretability of smart systems and provide explainable feedback to users. Moreover, some XAI methods can evaluate the model as independent from model-wise mechanisms. Some of these XAI methods are as follows:

- *SHAP*: An acronym for SHapley Additive exPlanations, SHAP uses Game Theory to understand how each input feature is effective in output(s). As developed by Lundberg and Lee (2017), SHAP basically considers each input feature as a player in the game, and it performs tries over the dataset (for training AI model) to understand how each input level affects the way of output(s). From this perspective, SHAP is a model-independent XAI method.
- *LIME*: Another model-independent XAI method, LIME (Local Interpretable Model-agnostic Explanations) derives synthetic data to understand how the target AI model is acting against this new data when compared with the original training data. In this way, it aims to understand how the designed AI model is safe and good enough for inferencing about the target problem. LIME can be effectively used for different types of data including raw text and images (Holzinger et al., 2020).
- *CAM*: CAM (Class Activation Mapping) is a XAI method that needs to be integrated into AI models that work with image-based data. Typically, CAM is used as a layer structure of neural networks to create heat maps over input image data. According to the resulting heat maps, it can be understood which parts of the input image was considered more by the model. In this way, any analysis problems of the model on image data can be detected easily (Poppi et al., 2021).
- *Interpretable Machine Learning (e.g., Decision Trees)*: Traditional machine learning models can be used for creating explainability for complex AI models. Since traditional machine learning algorithms are generally interpretable (having fewer parameters and easy to understand inside mechanisms), they can be integrated into advanced models to define output channels of explainability. For example, decision trees can be used for catching specific parameters and

designing decision-making trees, which can explain how the input-output relation is established by the related model (Mahbooba et al., 2021). In addition to decision trees, other machine learning algorithms such as Linear Regression and kNN can be used for the same purposes (Arrieta et al., 2020).

Considering the mentioned XAI methods, some samples of medical tourism scenarios can be defined directly. Since different types of data and complexity levels of data processing may take place in medical tourism ecosystems, XAI can be effectively used to improve the trust level. Table 10.1 provides some medical tourism scenarios by matching them with the XAI methods. The XAI methods listed in the table are used to improve the trustworthiness of the scenarios, by using special interfaces or outputs for the users. In this way, actual medical tourism scenarios are covered with safe algorithmic components. These scenarios may be smaller phases of wider applications.

10.4 DISCUSSION ON PRESENT STATE AND FUTURE

When the role of XAI in medical tourism is evaluated in terms of current state, it is possible to see that there are still more steps to take for deeper analysis of XAI methods specifically for medical tourism. However, by thinking about problem formulations for digitally improved medical healthcare approaches, we can discuss how advancements progress. Since collecting human data becomes practical through technological tools, more studies can be made about collaborations between hardware and software components. Mobile devices are good at user-experience and data-oriented tasks. Additionally, wearables have been gaining popularity. But all these devices need XAI usage at the software side to ensure a collaborative approach in the smart systems used in medical tourism applications. Use of XAI may include data collection from potential or active health tourists and therefore privacy should be ensured. Moreover, data errors must be considered. When it is considered from the perspective of service providers, XAI may improve the trust level of smart systems in terms of automated recommendations (e.g., detecting health tourists, grouping them, predicting intentions and costs, planning activities).

XAI can be more meaningful for medical staff. Medical staff including doctors may require trustworthy smart systems to help them plan medical operations and understand possible outcomes of treatments for health tourists. XAI is also vital for smart medical devices including robotic systems and digital diagnosis or screening machines.

XAI in medical tourism requires hierarchical use of XAI methods. This is because multiple smart systems and different human factors with different kinds of data are used inside the same environment. Use cases can be planned according to processed data as well as the AI techniques used. Simple but not transparent enough data processing algorithms may be supported with interpretable techniques such as decision trees. On the other hand, more advanced models such as artificial neural networks may use SHAP and LIME. In tasks for processing image data, CAM is be an effective XAI method for increasing the trust level. At the final stage, combinations of all these XAI-based analyses would be used to build the entire medical tourism plan.

TABLE 10.1

Some Scenarios of XAI Usage for Medical Tourism

Scenario	XAI Method
Health tourists' health states are tracked.	SHAP, LIME, Interpretable Machine Learning
Special promotions for target health tourists are automatically prepared.	SHAP, LIME, Interpretable Machine Learning
Accommodation venues and treatment clinics for health tourists are determined.	CAM (for visual data-based decision-making) SHAP, LIME, Interpretable Machine Learning
Health tourists communicate with chat bots.	LIME
Health tourists take photos for automated physical and emotional health state prediction.	CAM
Medical diagnosis and treatment routes for health tourists are determined.	SHAP, LIME, Interpretable Machine Learning
Medical images are diagnosed for doctors' decision making.	CAM
Doctors and medical staff receive decision support for check-ups, treatments, etc.	SHAP, LIME, Interpretable Machine Learning
Health tourists' physical rehabilitations are tracked and special treatments are planned.	CAM (for visual explainability) SHAP, LIME (for explainability of raw data)
Health tourists' medical treatment actions are tracked.	SHAP, LIME, Interpretable Machine Learning
Service providers manage resources and ongoing medical tourism applications.	SHAP, LIME, Interpretable Machine Learning
Service providers track medical tourism flows around the world, over a visual map.	CAM (for visual explainability) SHAP, LIME (for explainability of raw data)
Health tourists, doctors and service providers are matched automatically.	SHAP, LIME, Interpretable Machine Learning
Costs for medical tourism plans are checked.	SHAP, LIME, Interpretable Machine Learning
Decision support for business strategies and management are received.	SHAP, LIME, Interpretable Machine Learning
Success probability of medical tourism applications are predicted.	SHAP, LIME, Interpretable Machine Learning
Medical staff check and control medical devices, equipment, resources, etc., for sustainable, safe operations.	CAM (for devices aiming visual screening and diagnosis) SHAP, LIME, Interpretable Machine Learning (for explainability of outputs regarding non-image data)
Communication among all actors of a medical tourism application is tracked and managed	SHAP, LIME, Interpretable Machine Learning

XAI use in medical tourism scenarios may be connected to not only specific stages but also general management of the system. Since it is possible to develop a tracking platform for the tasks performed in a medical tourism program, XAI can be effectively used to create safe user interfaces. Thus, a health tourist can check if the applications are providing rational suggestions and ensuring correct and meaningful personalized costs, plans or promotions. For medical staff, this may be used for tracking operations, treatments, and safety of decision support by AI.

Since XAI-supported medical tourism applications may be innovative for improving service experience and revenue, some regulations should be put in place for XAI usage. This may be connected to other stakeholders when both technical and business side of the health tourism area is considered widely.

10.5   CHALLENGES AND OUTLOOK

Although use cases of XAI for medical tourism are open, there are some challenges related to users or technological limitations in terms of XAI design and combining the design with medical tourism plans. These challenges are briefly as follows (Figure 10.3):

- **Integration**: Although XAI provides flexibility for explainability, it may be difficult to integrate multiple XAI methods in advanced medical tourism applications with many users. Such situations may require careful analysis and integration practices by experts and stakeholders.
- **Costs**: Associated with globally wide medical tourism ecosystems, adding XAI methods or building XAI-based smart systems may cause increased costs for renewal of data systems or purchasing new technological components to run XAI.
- **Technology Acceptance**: Health tourism is affected by technological changes. It has been digitally transformed as a result of advancing information and communication tools. However, service providers, managers and stakeholders may stay away from smart systems and XAI. So, it is necessary to increase the awareness towards safe, sustainable, trustworthy AI.
- **Regulations:** Since medical tourism applications are associated with policies and rules determined by host countries, some use cases of XAI may encounter restrictions in integrating external components with private data. In such cases, global regulations for medical tourism should be extended for use of advanced technology such as XAI.

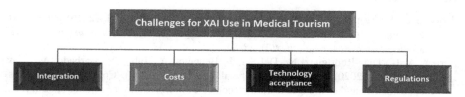

FIGURE 10.3   Challenges regarding XAI use in medical tourism.

Connected with discussions regarding the present state of XAI and medical tourism, some future predictions may be made:

- In the future, usage of XAI may be a common practice or simply a requirement for digital components of future medical tourism (or generally health tourism) applications.
- In the future, there will be more mobile devices, wearables, and special software systems to collect human data. Thus, the future of medical tourism applications will require more XAI usage to increase acceptance and trust.
- In the future, more types of XAI will be available to be integrated into medical tourism applications. In some cases, XAI-supported software systems will come with medical devices.
- Medical tourism applications will expand more and more in the future. As a result of this, XAI will be a common, end-user component for health tourists to check instantly how their data is shared with service providers and how they process it to provide accurate medical tourism services.
- In the future, use of XAI may affect not only health tourists, medical staff, and service providers but also policymakers or regulators around the world.

10.6   CONCLUSIONS

This chapter provided an overview of XAI aspects with regard to medical tourism. Medical tourism is in a remarkable place to benefit from current advancements since data processing solutions tend to extend the capabilities in medical tourism applications. As discussed, smart systems can extend medical tourism for better communication of actors (health tourists, medical staff, service providers) and accurate data processing. However, it is a necessity to think about the potential use of XAI so that trustworthy systems for medical tourism applications can be developed accordingly. XAI methods should ensure data integrity and trust among users. These methods may include SHAP, LIME, CAM, and even integrated interpretable machine learning algorithms to build the desired XAI infrastructure. These methods are vital when multiple smart systems are used for establishing a big application ecosystem for medical tourism scenarios. Future research should focus on design and development of ecosystems, human-based evaluations of XAI, and having feedback from the actors of health tourism in order to understand how the existence of XAI affects the development of trustworthy, sustainable applications.

REFERENCES

Amouzagar, S., Mojaradi, Z., Izanloo, A., Beikzadeh, S., & Milani, M. (2016). Qualitative examination of health tourism and its challenges. *International Journal of Travel Medicine and Global Health*, *4*(3), 88–91.

Arrieta, A. B., Díaz-Rodríguez, N., Del Ser, J., Bennetot, A., Tabik, S., Barbado, A., ... & Herrera, F. (2020). Explainable Artificial Intelligence (XAI): Concepts, taxonomies, opportunities and challenges toward responsible AI. *Information Fusion*, *58*, 82–115.

Aydin, G., & Karamehmet, B. (2017). Factors affecting health tourism and international health-care facility choice. *International Journal of Pharmaceutical and Healthcare Marketing, 11*(1), 16–36.

Confalonieri, R., Coba, L., Wagner, B., & Besold, T. R. (2021). A historical perspective of explainable Artificial Intelligence. *Wiley Interdisciplinary Reviews: Data Mining and Knowledge Discovery, 11*(1), e1391.

Connell, J. (2006). Medical tourism: Sea, sun, sand and... surgery. *Tourism Management, 27*(6), 1093–1100.

Druce, J., Harradon, M., & Tittle, J. (2021). Explainable artificial intelligence (XAI) for increasing user trust in deep reinforcement learning driven autonomous systems. *arXiv preprint arXiv:2106.03775.*

Gomes, C. P. (2000). Artificial intelligence and operations research: Challenges and opportunities in planning and scheduling. *Knowledge Engineering Review, 15*(1), 1–10.

Guo, W. (2020). Explainable artificial intelligence for 6G: Improving trust between human and machine. *IEEE Communications Magazine, 58*(6), 39–45.

Holzinger, A., Saranti, A., Molnar, C., Biecek, P., & Samek, W. (2020). Explainable AI methods- a brief overview. In: *International Workshop on Extending Explainable AI Beyond Deep Models and Classifiers* (pp. 13–38). Cham: Springer International Publishing.

Hopgood, A. A. (2021). *Intelligent systems for engineers and scientists: a practical guide to artificial intelligence.* Boca Raton, FL, USA: CRC Press.

Isa, S. M., Ismail, H. N., & Fuza, Z. I. M. (2020). Elderly and heritage tourism: A review. In: *IOP Conference Series: Earth and Environmental Science* (Vol. 447, No. 1, p. 012038). Bristol, UK: IOP Publishing.

Jackson, P. C. (2019). *Introduction to artificial intelligence.* New York: Courier Dover Publications.

Jiang, L., Wu, Z., Xu, X., Zhan, Y., Jin, X., Wang, L., & Qiu, Y. (2021). Opportunities and challenges of artificial intelligence in the medical field: current application, emerging problems, and problem-solving strategies. *Journal of International Medical Research, 49*(3), 03000605211000157.

Joshi, A. V. (2020). *Machine learning and artificial intelligence.* Cham: Springer International Publishing.

Khosravi, H., Shum, S. B., Chen, G., Conati, C., Tsai, Y. S., Kay, J., ... & Gašević, D. (2022). Explainable artificial intelligence in education. *Computers and Education: Artificial Intelligence, 3,* 100074.

Kilavuz, E. (2018). Medical tourism competition: The case of Turkey. *International Journal of Health Management and Tourism, 3*(1), 42–58.

Loh, H. W., Ooi, C. P., Seoni, S., Barua, P. D., Molinari, F., & Acharya, U. R. (2022). Application of explainable artificial intelligence for healthcare: A systematic review of the last decade (2011–2022). *Computer Methods and Programs in Biomedicine,* 107161.

Lundberg, S. M., & Lee, S. I. (2017). A unified approach to interpreting model predictions. In: *Proceedings of the 31st International Conference on Neural Information Processing Systems* (pp. 4768–4777), December 4–9, Long Beach, CA, USA.

Mahbooba, B., Timilsina, M., Sahal, R., & Serrano, M. (2021). Explainable artificial intelligence (XAI) to enhance trust management in intrusion detection systems using decision tree model. *Complexity, 2021,* 1–11.

Majeed, S., Lu, C., Majeed, M., & Shahid, M. N. (2018). Health resorts and multi-textured perceptions of international health tourists. *Sustainability, 10*(4), 1063.

Mehdiyev, N., & Fettke, P. (2021). Explainable artificial intelligence for process mining: A general overview and application of a novel local explanation approach for predictive process monitoring. In: W. Pedrycz, & S.-M. Chen (Eds.), *Interpretable Artificial Intelligence: A Perspective of Granular Computing* (pp. 1–28). Springer, Cham.

Michalski, R. S., Carbonell, J. G., & Mitchell, T. M. (Eds.). (2013). *Machine learning: An artificial intelligence approach.* Berlin, Germany: Springer Science & Business Media.

Moghavvemi, S., Mogan, K., & Ghazali, E. M. (2023). The issue, challenges and risk of post-surgery treatment abroad among medical tourists from doctors' perspective. *Journal of Quality Assurance in Hospitality & Tourism*, 1–32.

Mohanty, A., & Mishra, S. (2022). A comprehensive study of explainable artificial intelligence in healthcare. In: S. Mishra, H. K. Tripathy, P. Mallick, & K. Shaalan (Eds.), *Augmented Intelligence in Healthcare: A Pragmatic and Integrated Analysis* (pp. 475–502). Singapore: Springer Nature Singapore.

Nisa, H., & Sharma, S. K. (2023). Status and growth of medical tourism In India. *Journal of Namibian Studies: History Politics Culture*, *35*, 5004–5011.

Ormond, M. (2014). *Medical tourism.* The Wiley-Blackwell Companion to Tourism. Hoboken, NJ, USA: Wiley.

Pannu, A. (2015). Artificial intelligence and its application in different areas. *Artificial Intelligence*, *4*(10), 79–84.

Poppi, S., Cornia, M., Baraldi, L., & Cucchiara, R. (2021). Revisiting the evaluation of class activation mapping for explainability: A novel metric and experimental analysis. In *Proceedings of the IEEE/CVF Conference on Computer Vision and Pattern Recognition* (pp. 2299–2304), June 19–25, Nashville, TN, USA.

Reisman, D. A. (2010). *Health tourism: Social welfare through international trade.* Edward Elgar Publishing.

Sag, I., & Sengul, F. D. (2019). Why medical tourists choose Turkey as a medical tourism destination? *Journal of Hospitality and Tourism Insights*, *2*(3), 296–306.

Smith, M., & Puczkó, L. (2014). *Health, tourism and hospitality: Spas, wellness and medical travel.* London, UK: Routledge.

11 XAI for Advancements in Drug Discovery

Waseem Ullah Jan, Naseer Ali Shah, and Haroon Ahmed

11.1 INTRODUCTION

Explainable artificial intelligence (XAI) is a programmable artificial intelligence with the ability to describe the purpose of its creation, reasoning, and decision-making process. This description is generated in the most basic form, understandable for a layman of the field (Wigmore, 2018). XAI can be termed as a powerful tool with the ability to answer the *hows* and the *whys*. It can also be used to address ethical and legal issues related to artificial intelligence (AI) (Turri, 2022). XAI is a tool used for evaluating effectiveness and explanability of different AI models. It proposes different machine learning techniques (ML) that enable creation of AI models that are explainable, without compromising on learning performance. Furthermore, these techniques should also enable humans, in terms of understanding, trusting, and managing AI models (Gunning, 2017)

11.1.1 BACKGROUND AND HISTORY

Historically, the first AI models that was self-explanatory were decision trees developed in 1990 (Quinlan, 1990). Other early models of AI that were easily interpretable included experts systems, fuzzy logics, and automated reasoning (Robinson & Voronkov, 2001). Artificial neural networks (ANNs) were also more explicable as compared to recent AI models. Most prominent models of such ANNs include radial-basis function (RBF) architecture and linguistic (Hearst, Dumais, Osuna, Platt, & Scholkopf, 1998).

ML is a technique used to teach data handling to machines (software). It can be considered a branch of AI. ML allows interpretation of data which is not possible by a human. In ML, first different arbitrary or real datasets are fed to machines to make them capable of interpretation. Then, the data under research is provided to the machine which interprets it (Jordan & Mitchell, 2015; Mahesh, 2020). ML consists of layers with each layer accepting an input, processing it, and passing it to next layer. If an ML algorithm is able to process and interpret data using three different layers, it is called deep learning (DL). Hence DL can be termed as a sub-set or branch of ML. It is a neural network that aims to mimic the human brain while learning from large datasets (LeCun, Bengio, & Hinton, 2015; Kamilaris & Prenafeta-Boldú, 2018).

DOI: 10.1201/9781003426073-11

Recently, a shift of focus from accuracy to explainability has been observed. Researchers are now looking at opening the black box and searching for alternatives for them (Angelov & Soares, 2020). This is where the need of XAI arises. XAI models can be explained in terms of transparency, with the potential to be understandable interpretability, with the ability to provide explanations that are comprehensible for humans; and explainability, the belief that explanation serve as an connection between the human and the AI system (Adadi & Berrada, 2018; Gilpin et al., 2018)

In 2018, the National Institute of Standards and Technology (NIST) published a report in which the four principles of XAI were explained. Explanation is the first principle. According to this principle, a system must be able to provide evidence, arguments, and/or reasoning for produced results regardless of whether those arguments and/or reasoning is correct or not. The explanation must be meaningful and understandable by users. This is the second principle. The third principle requires the explanation and its meaning to be accurate and precise. Knowing limits of jurisdiction and/or knowledge is the last principle. This enables the developer to maintain operations of XAI within the limits prescribed by different regulatory authorities (Phillips, Hahn, Fontana, Broniatowski, & Przybocki, 2020).

The aim of this chapter is to provide insights into the mechanisms behind XAI. It will also elaborate on how DL has been used in drug discovery as well as the different models of DL in drug discovery. Furthermore, this chapter explains the different models of XAI that have been used for various purposes. In addition to this, aspects as well as limitations of XAI are discussed. Lastly, the drugs that have been discovered using XAI are mentioned.

11.2 DEEP LEARNING TECHNIQUES

11.2.1 NEURAL NETWORKS

The development of neural networks was inspired by the human brain. They are composed of nodes that are interconnected and function like a neuron system to process as well as analyze data. They can also be used to make predictions and decisions. Many different categories of neural networks exist (Nielsen, 2015), but here we only discuss those used in drug discovery.

11.2.1.1 Classical Neural Networks

Classical neural networks are also referred to as artificial neural networks (ANNs) and either allow flow of information is or data in one direction only, i.e. from input to output. Data is only converted from input to output. The layers in classical neural networks include an input layer, hidden or intermediate layers, and an output layer. The input layer receives the data followed by their processing by the hidden layers and then the output layer produces the results (Mukhamediev, Symagulov, Kuchin, Yakunin, & Yelis, 2011).

11.2.1.2 Convolutional Neural Networks

Convolutional neural networks (CNNs) are advanced ANNs. They were developed to manage complex data and allow pre-treatment as well as compilation. They are the most flexible neural networks. They can take input in the form of text as well as images. CNN layers differ in accordance with their function. When analyzing basic visual data, only one input layer is used. In some CNNS, output has a single dimensional layer while others may have an extra sampling layer that restricts the number of neurons involved in processing. Lastly, in some CNNs, connected layers are used to join sampling and output layers (Yamashita, Nishio, Do, & Togashi, 2018)

11.2.1.3 Recurrent Neural Networks

They were initially created for sequence prediction. These networks require input with different lengths. Recurrent neural networks (RNNs) forecast expected results on the basis of knowledge of input with variable lengths. RNNs use long short-term memory (LSTM) algorithms that allow retention of short time information (input) over a long period of time. LSTMs rely on input, output, and forget gates (Chandra & Sharma., 2017; Zia & Zahid, 2019).

11.2.1.4 Generative Adversarial Networks

Generative adversarial networks (GANs) combine two different neural network approaches. This type of network combines a generator, which is used for generating images, while the discriminator allows differentiation between real and fake images (Hsu, Zhuang, & Lee, 2020). Both of these networks are in competition with each other. The generator keeps on generating fake data while the discriminator has to differentiate between the real and fake data (Aggarwal, Mittal, & Battineni, 2021).

11.2.2 SELF-ORGANIZING MAPS

Different types of maps have been used in XAI as well as DL. However, self-organizing maps (SOMs) are of particular importance. These maps utilize the unsupervised data in order to reduce random variables of a model. This approach has

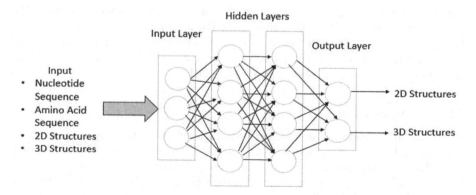

FIGURE 11.1 Simplified workflow of classical and convolution neural network (modified from Mukhamediev, Symagulov, Kuchin, Yakunin, & Yelis, 2011; Yamashita, Nishio, Do, & Togashi, 2018)

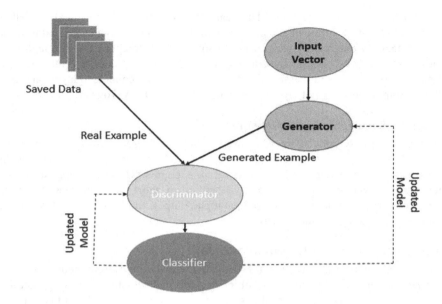

FIGURE 11.2 Simplified workflow of a Generative Adversarial Network (modified from Aggarwal, Mittal, & Battineni, 2021).

2D output dimensions. This allows the weight of best matching units to be adjusted. The value of weights of each BMU varies because weight is the attribute of the node itself (Askr, et al., 2022).

11.2.3 BOLTZMANN MACHINE

These neural networks use algorithms that produce results that are randomly distributed but not precisely predicted. These networks create a binary recommendations platform for analyzing data. They have two layers. The first one is called a visible or input layer while the second is called a hidden one. The nodes in a Boltzmann machine are interconnected across levels but not in the same layers. This results in disconnection among layers, which is a disadvantage (Hinton, 2012).

11.2.4 AUTOENCODERS

These neural networks allow unsupervised processing of data. They learn to compress as well as reconstruct input data. They have two components, an encoder that maps the input data into a lower dimension representation and a decoder that works in the opposite way. Various types of autoencoders exist. The sparse autoencoders have more hidden layers than input layers. This hinders autoencoders to use all nodes simultaneously while contractive ones also have the same structure as sparse autoencoders except they introduce a penalty factor to the function that is lost. Deionizing autoencoders randomly equal the data to zero while stack autoencoders have an extra hidden layer that allows double encoding (Zhai, Zhang, Chen, & He, 2018).

11.3 DEEP LEARNING IN DRUG DISCOVERY

11.3.1 DRUG-TARGET INTERACTION (DTI) PREDICTION

DTI prediction is the first step in creating new medications. It is also one of the most crucial steps in screening as well as drug-guided synthesis. There are more many approaches that can be used while predicting DTIs (Wang et al., 2020). These include the following.

11.3.1.1 Drug-Based Models

These models are based on the hypothesis that a particular drug will be similar to known drugs in physical and/or chemical properties. Structural similarity is a key factor in these models (Thafar, Raies, Albaradei, Essack, & Bajic, 2019). Many different models can be used. CNN is a common technique. Molecular descriptor-based models can allow ADMET (Adsorption, Distribution, Metabolism, Excretion and Toxicity) analysis. Some CNNs are based on atomic donor-acceptor pairs while others rely on pharmacophoric donor-acceptor pairs. SMILES have also been used by CNNs. GCNs can be trained to use 3D descriptors (Askr et al., 2022).

11.3.1.2 Structure-Based Models

These models rely on the structure of both protein and drug. Typically, they include molecular docking and simulations (Askr et al., 2022). However, different researchers have trained different CNNs for predicting DTIs based on structural similarity. DeepConv-DTI used CNNs (Lee, Keum, & Nam, 2019). Apart from CNNs and GCNs, Extended-Connectivity Fingerprints (ECFPs) have also been used as pharmacological features in structure-based models. ECFPs are topological fingerprints that are used for molecular characterization (Feng, Dueva, Cherkasov, & Ester, 2018). Additionally, LSTMs and RNNs have also been utilized for prediction of protein sequences as well as their docking (Gao et al., 2017).

11.3.1.3 Drug-Protein-Based Models

This approach is particularly useful for drug selectivity or protein promiscuity. In addition to this, drug dosage and effects on primary as well as secondary targets can also be identified through this approach (Cortés-Ciriano et al., 2015). Several DL technologies have been designed for this approach. Drug-induced gene expression patterns and DTI-related heterogenous networks are two common DL approaches. The heterogeneous networks use a variety of nodes and edge kinds (Luo et al., 2017).

11.3.2 DRUG SENSITIVITY AND RESPONSE PREDICTION

This involves predictions of drugs that are more effective in treating a particular disease. It also allows prediction of a person or proteins sensitivity to a particular drug as well as multiple drugs (Askr et al., 2022). DNNs are the most commonly used DL technique for this approach. Previous DNNs, poor generalization as well as overparameterization have made them unfit for this approach. However, recently created DNNs have shown promising results (Gómez-Bombarelli et al., 2018). Apart from DNNs, CNNs have also been used in this approach. CDRscan relies on CNNs, and it can predict 1000 reactions per molecule (Chang et al., 2018).

Autoencoders can also be used to predict response of drugs. DeepProfile is an autoencder-based DL technique that allows prediction of gene expression in Acute Myeloid Leukemia (AML) patients (Dincer, Celik, Hiranuma, & Lee, 2018). Similarly, another autoencoder was proposed that allows learning of gene expression in cancer cells. It also predicted copy number variations and somatic mutations (Ding, Chen, Cooper, Young, & Lu, 2018). ANNs and GNNs can also be used for prediction of drug response (Askr et al., 2022). Twin CNNs for drugs in SMILES format (TCNNS) was coded for representation of drugs as well as feature vectors of cell lines. It has two subnetworks, each encoding one dimensional input data (Cortés-Ciriano et al., 2015).

Multi-omics Late Integration (MOLI) is a DL technique launched in 2019. It has the ability to characterize cell lines by incorporation of somatic mutations as well as multi-omics data. There are three subnetworks in MOLI, each for specific omics data. The cell's response is identified or generated via a final network. MOLI allows incorporation of multiple input data as well as binary classification of data response (Sharifi-Noghabi, Zolotareva, Collins, & Ester, 2019).

11.3.3 DRUG-DRUG INTERACTION (DDI) SIDE-EFFECT PREDICTION

Drugs have strong affinity to bind to their target proteins. However, they also have weak affinity to bind to various non-target proteins that may result in side effects (Liu et al., 2022). Prediction of DDIs can result in reduction of side reactions as well as optimization of drug dosage and development. The side effects and/or reaction can often result in failure of drug development. Therefore, there is a need for development of DL techniques for prediction of DDIs (Askr et al., 2022).

One of the earliest DL techniques for DDI prediction was Tiresias. It was developed in 2016. Tiresias takes input in the form of large data that is segmentally integrated. It establishes a knowledge network that is representative of properties of drugs and its various interactions with components. It uses enzymes, chemical and physical structure, and route of intake for formation of network. For DDI prediction, a logistic regression prediction model is utilized (Ferdousi, Safdari, & Omidi., 2016).

Relational GNNs can be used for DDI prediction. This model uses the chemical structure of a drug as input. DDIs are predicted by either implementing a multitasking learning prediction model or vien Decagon polypharmacy side effects model (Ryu, Kim, & Lee, 2018; Zitnik, Agrawal, & Leskovec, 2018). Apart from GNNs, autoencoders can also be used in DDI side-effect prediction. An autoencoder factorizes intricate non-linear pharmacological interactions for their better understanding (Chu et al., 2018).

11.4 MODELS OF XAI

Most of the studies on ML and XAI are based on their sensitivity analysis. The feature relevance explanation method is a common post-hoc text that can be used during sensitivity analysis. This method utilizes a scoring method to compare different variables and determine the importance given to each variable while generating output. This post-hoc test is commonly needed for tree ensembles, support vector machines, multi-layer, conventional, and recurrent neural networks (Arrieta et al., 2020). Local Interpretable Model-Agnostic Explanations (LIMEs) have also been studied in detail. This framework allows explaining the image output generated by XAI. However, its, tabular explainer has rarely been discussed and analyzed. Being local means analyzing specific observations and/or variables rather than providing a general explanation of the whole model. LIMEs provide information on about how a particular portion of an image affect the generated result (decision). Lastly, it can be applied to any black-box algorithm that has ever been generated or written (Dieber & Kirrane, 2020).

11.4.1 Feature-Oriented Models

SHaply Additive explanation (SHAP) is a commonly used ML prediction method. It is based on a game-theoretic approach. In this approach, each feature or variable is assumed as a player in a coalition game. The ultimate result or explanation is based on the combined performance of each variable. The model works as an additive function with each variable adding its output into the model output. However, if this additive function is disturbed, the transparency of results become compromised (Angelov P. P., Soares, Jiang, Arnold, & Atkinson, 2021). SHAP provides flexibility while selecting and evaluating models. This flexibility allows comparison of various frameworks under the same metrics. However, it has been reported that SHAP implementation cannot be optimized efficiently for each and every model (Lundberg & Lee, 2017)

The second most important class of features-oriented models are class activation maps (CAMs). These models allow representation of linear sums of visual patterns. These patterns are located at different spatial positions. The linear sum is always weighted per class (Angelov P. P., Soares, Jiang, Arnold, & Atkinson, 2021). CAMs use global average pooling to weigh the linear sum and visualize them through a heat map. CAMs are mostly applied to CNNs, as both CAMs and CNNs learn directly from the data instead of being pre-trained. Furthermore, it has been reported that using fully connected layers and map scaling can result in loss of spatial information (Zhou, Khosla, Lapedriza, Oliva, & Torralba, 2016).

Gradient-weighted-CAM (Grad-CAM) and Grad-CAM[++] use CAM as a base model (Angelov P. P., Soares, Jiang, Arnold, & Atkinson, 2021). Grad-CAM generalizes the CAM model to any CNN that is arbitrary. This allows computation of important scores and a heatmap indicates regions of input that were more important to the generated output. However, it has been reported that Grad-CAM cannot explain the use of the same object from an image at multiple instances. Furthermore, the visualizations produced by Grad-CAM are coarse-grained only (Selvaraju et al.,

2017). To overcome these drawbacks. Grad-CAM++ was developed. It considers weighted average instead of linear sum (Chattopadhay, Sarkar, Howlader, & Balasubramanian, 2018).

11.4.2 GLOBAL MODELS

Global attribution mappings (GAMs) allow explanation of global-level predictions of neural networks. GAMs calculate attributes as weighted combined rankings. Data from different sub-populations can be interpreted through an adjustable condition or level. GAMs clusters different local features by finding pair-wise distance matrixes. GAMs allow its user to study common features among different sub-populations (Ibrahim, Louie, Modarres, & Paisley, 2019; Wu et al., 2020).

Another visualization model of XAI is gradient-based saliency maps (GSPs). These maps can provide absolute values of gradient in the form of normalized heat map. These values are provided with respect to input features. High activity areas are presented in the form of highlighted pixels. This allows users to look at features that are most influential. However, if gradients of neurons have a negative value it would be suppressed as the values are absolute (Simonyan, Vedaldi, & Zisserman, 2014; Mahendran & Vedaldi, 2016).

Deep attributed maps (DAPs) are a technique used for generating explanation of gradient-based maps such as GSPs. DAPs can provide a framework that has the ability to evaluate different saliency-based models. The output of DAPs is in the form of heatmaps. It is generated by multiple gradient output with its respective input. DAPs can determine whether the input point has a positive or negative contribution. However, noisy gradients and variation in input can affect explanation. Furthermore, if two models produce similar or different results, DAPs would not be able to explain the logic behind those results (Ancona, Ceolini, Öztireli, & Gross, 2017, 2019).

11.4.3 CONCEPT MODELS

Conception activation vectors (CAVs) is a technique that explains the internal states of a neural network globally. They map human understandable features to features that are latent. These latent features are extracted by neural networks (Kim et al., 2018). Based on extraction of CAVs, automatic concept-based explanation models are built. These models allow segmentation of human understandable concepts at various spatial resolutions. These models rely on uniqueness as well as the effectiveness of explanation of concepts (Ghorbani, Wexler, Zou, & Kim, 2019).

11.4.4 SURROGATE MODELS

These models are used when no direct outcome can be predicted or explained. The LIME technique can be used for any type of model and is used for training an interpretable surrogate model. This technique allows surrogate models to learn the local behavior of black box which allows the global prediction. This means that a surrogate model would allow us to interpret prediction of the black box at each pixel of an image (Dieber & Kirrane, 2020).

11.4.5 LOCAL PIXEL-BASED MODELS

Many different algorithms exist for providing explanation of each pixel of a ML-generated image. Layer-wise relevance propagation (LRP) relies on pre-defined propagation rules. It can provide explanation of output from a multi-layered neural network. LRP generates a heatmap that signifies the contribution of each pixel in the prediction of output; LRP is a post-hoc test and can only be applied if the neural network allows backpropagation (Bach et al., 2015). DeconvNet relies on semantic segmentation and provides explanations about the contributions of each pixel (Noh, Hong, & Han, 2015). Lastly, a deep belief network was proposed that can enhance the interpretability of CNNs (Hinton et al., 2006).

11.4.6 HUMAN-CENTRIC MODELS

A new model was proposed that differs from all previously explained models. The basic advantage of this new model is its human centric (anthropomorphic) nature. This model allows the production of clear and human-understandable explanations, and provides explanation of an output as a whole image and/or audio rather than a per-pixel explanation (Angelov & Soares, 2020).

11.5 XAI IN DRUG DISCOVERY

Computer-aided drug design (CADD) technologies have been widely used in drug discovery. They reduce the number of screening tests as well as the number of ligands that need to be screened. CADD technologies also reduce the cost of operation while ensuring prediction of the best possible lead compounds (Leelananda & Lindert, 2016). XAI is new in the drug discovery field. However, it is developing at a rapid pace (Jiménez-Luna, Grisoni, & Schneider, 2020).

Various XAI models have been used in drug discovery. GAMs have been used to detect ligand pharmacophores. Sixteen binary label sets were constructed. Binding was defined by the presence or absence of a particular label. Message passing neural networks (MPNNs) were used as a model for binding. MPNNs attribute binding score, in the form of 0 or 1, to each individual atom. GAMs were used to evaluate and explain assignment of those binding scores. The performance of GAMs at explaining prediction of model was satisfactory. However, there was a need to learn about spurious correlation (McCloskey, Taly, Monti, Brenner, & Colwell, 2019). CAMs can be applied to CNNs. However, they cannot be readily applied to graph-CNNs (GCNNs). For this reason, Grad-CAM and contrastive excitation backprop (EB) may be used. This model can be effective for identifying functional groups present on ligands when determining adverse effect of a lead compound (Pope, Kolouri, Rostami, Martin, & Hoffman, 2019).

Structural-activity relation (SAR) analysis is one of the major components of CADD. ML has significantly improved prediction of activity of a lead compound through various technologies (Rodríguez-Pérez & Bajorath, 2019). Support vector machines (SVMs) can be used to perform multi-class classification as well as regression analysis (Pisner & Schnyer, 2020). Regression analysis is necessary while performing SAR analysis. SHAP can be used to compare predicted models from different SVMs and DLs to provide better characterization of protein-ligand interaction. It can also be used for predicting multi-target activity simultaneously (Rodríguez-Pérez & Bajorath, 2019). Feature attribution models, such as SHAP, can also be used for explaining results of root mean square deviation (RMSD: a measure for the difference between the predicted result and the actual result) values as well as their comparison and pose distinguishing based on RMSD values. Similarly, this model also allows affinity prediction of ligands as well as proteins (Hochuli, Helbling, Skaist, Ragoza, & Koes, 2018).

Gradient-based models such as Grad-CAM, Grad-CAM^{++}, and GSPs can be used in DNN-based predictions. Various black-box DNN models employ these gradient-based maps and models. These DNN models can be used for DDI prediction (Sun, Ma, Du, Feng, & Dong, 2018). In addition to DNN models, DeepLIFT can be used to top the results produced from DNN models. They have been shown to be superior techniques. DeepLIFT is also based on gradient-based models (Shrikumar, Greenside, & Kundaje, 2017). Another technique breaks down input into small text. This technique utilizes GANs, which contains a a generator and an encoder. As this model is similar to other natural language processing (NLP) models, it may be used for prediction of DDIs (Lei, Barzilay, & Jaakkola, 2016).

Instance-based approaches have not been utilized in drug discovery. However, it is possible to utilize these approaches during cliff prediction due to their ability to predict minor changes in structure of molecules during biological activity. Similarly, they have the ability to highlight the minimum number of atoms responsible for a particular activity, also known as fragment-based screening. Lastly, they may allow predictions of minimum changes required in the structure of a drug to enhance biological activity as well as physiochemical properties (Jiménez-Luna, Grisoni, & Schneider, 2020).

GNNs are gaining attention rapidly and there is a need for enhancing their explanations. GNNExplainer was developed to explain results obtained via GNNs. It relies on mask optimization for learning soft masking for nodes and edge attributes. These masks are then compared with initial results of GNNs to provide changing in formation (Ying, Bourgeois, You, Zitnik, & Leskovec, 2019). GNNExplainer has been tested for predicting mutagenic functional groups present in *Salmonella typhimurium* (Debnath, Compadre, Debnath, Shusterman, & Hansch, 1991). Before these masks are tuned to input graphs, they may further be elaborated on using PGExplainer. It provides the approximate discrete edge masks. PGExplainer relies on a mask predictor that concatenates node embeddings to obtain embeddings of each edge of the initial input graph. This is followed by calculated prediction of each edge selection chances (Luo et al., 2020). GraphMask also elaborates on the results of GNNs. It explains the relevancy of each edge and forecasts which edge may be eliminated as it does not contribute to the final result (Schlichtkrull, Cao, & Titov, 2020).

XAI can provide graphical explanations of molecular interactions between drugs and targets. Cytochrome P450 (3A4 isoform, CYP3A4) interactions with drugs have been studied in detail. Gradient feature attribution models such as Grad-CAM and Grad-CAMM^{++} were combined with GCNNs to predict interactions. XAI was able to identify the chemical sub-structures of CYP3A4 that are involved in biotransformation

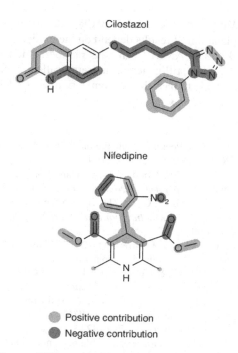

FIGURE 11.3 Cytochrome P450 metabolic sites (Deduced from Jiménez-Luna, Grisoni, & Schneider, 2020).

of drugs. Two of them are present in cilostazol; one is a tetrazole moiety, the other is the secondary amino group, while the second ones are the methyl and ester groups, which are metabolically labile (Jiménez-Luna, Grisoni, & Schneider, 2020).

11.6 LIMITATIONS OF XAI

One of the most prominent limitations of XAI is the inability of the human mind to understand and utilize the explanations derived via XIA for decision making. Secondly, there is a lack of quantitative measures to evaluate the accuracy and completeness of XAI-derived maps (Das & Rad, 2020). It has been observed that LIME-generated attributions are not always understandable (Mohseni & Ragan, 2018). One of the major limitations of feature-based models is their dependence on model input. The explanations provided by these models are dependent upon the original set of features. Another drawback is the use of complex molecular descriptors in input. Using complex descriptors opaques the model input and affects the interpretability of feature attribution models (Jiménez-Luna, Grisoni, & Schneider, 2020).

It has also been observed that small disruptions in input can lead to major changes in the output's explanation. This reduces the accuracy of the saliency maps generated by different models such as Deep LIFT and GSPs. Furthermore, biasness in input selection can also result in reduction of accuracy of these maps (Ghorbani, Wexler,

Zou, & Kim, 2019; Wang, Zhou, & Bilmes, 2019). Additionally, it has been reported that the result and explanation of results depend on the model and the algorithm used. Both Grad-CAM and Grad-CAMM++ can pass randomization tests. This means if data is randomized, they will still generate the same saliency maps. However, other XAI models will not be able to pass the randomization test (Adebayo et al., 2018).

REFERENCES

Adadi, A., & Berrada, M. (2018). Peeking inside the black-box: A survey on explainable artificial intelligence (XAI). *IEEE*, 6, 52138–52160.

Adebayo, J., Gilmer, J., Muelly, M., Goodfellow, I., Hardt, M., & Kim, B. (2018). Sanity checks for saliency maps. *Advances in Neural Information*, 31, 9505–9515.

Aggarwal, A., Mittal, M., & Battineni, G. (2021). Generative adversarial network: An overview of theory and applications. *International Journal of Information Management Data Insights*, 1(1), 100004.

Ancona, M., Ceolini, E., Öztireli, C., & Gross, M. (2017). Towards better understanding of gradient-based attribution methods for deep neural networks. *arXiv preprint arXiv*.

Ancona, M., Ceolini, E., Öztireli, C., & Gross, M. (2019). Gradient-based attribution methods. *Explainable AI: Interpreting, Explaining and Visualizing Deep Learning*, 169–191.

Angelov, P. P., Soares, E. A., Jiang, R., Arnold, N. I., & Atkinson, P. M. (2021). Explainable artificial intelligence: an analytical review. *Wiley Interdisciplinary Reviews: Data Mining and Knowledge Discovery*, 11(5), e1424.

Angelov, P., & Soares, E. (2020). Towards explainable deep neural networks (xDNN). *Neural Networks*, 130, 185–194.

Arrieta, A. B., Díaz-Rodríguez, N., Ser, J. D., Bennetot, A., Tabik, S., Barbado, A., … Herrera, F. (2020). Explainable Artificial Intelligence (XAI): Concepts, taxonomies, opportunities and challenges toward responsible AI. *Information Fusion*, 58, 82–115.

Askr, H., Elgeldawi, E., Ella, H. A., Elshaier, Y. A., Gomaa, M. M., & Hassanien, A. E. (2022). Deep learning in drug discovery: an integrative review and future challenges. *Artificial Intelligence Review*, 56(7), 5975–6037.

Bach, S., Binder, A., Montavon, G., Klauschen, F., Müller, K. R., & Samek, W. (2015). On pixel-wise explanations for non-linear classifier decisions by layer-wise relevance propagation. *PLoS One*, 10(7), e0130140.

Chandra, B., & Sharma., R. K. (2017). On improving recurrent neural network for image classification. *International Joint Conference on Neural Networks (IJCNN)* Anchorage (AK), USA, (pp. 1904–1907). IEEE.

Chang, Y ., Park, H., Yang, H. J., Lee, S., Lee, K. Y., Kim, T. S., … M., S. J. (2018). Cancer drug response profile scan (CDRscan): a deep learning model that predicts drug effectiveness from cancer genomic signature. *Science*, 8(1), 1–11.

Chattopadhay, A., Sarkar, A., Howlader, P., & Balasubramanian, V. N. (2018). Grad-cam++ : Generalized gradient-based visual explanations for deep convolutional networks. *In 2018 IEEE Winter Conference on Applications of Computer Vision (WACV)* Lake Tahoe, NV, USA, (pp. 839–847). IEEE.

Chu, X., Lin, Y., Gao, J., Wang, J., Wang, Y., & Wang, L. (2018). Multi-label robust factorization autoencoder and its application in predicting drug–drug interactions. arXiv:1811.00208.

Cortés-Ciriano, I., Ain, Q. U., Subramanian, V., Lenselink, E. B., Méndez-Lucio, O., IJzerman, A. P., … Bender, A. (2015). Polypharmacology modelling using proteochemometrics (PCM): recent methodological developments, applications to target families, and future prospects. *MedChemComm*, 6(1), 24–50.

Das, A., & Rad, P. (2020). Opportunities and challenges in explainable Artificial Intelligence (XAI): a survey. *arXiv:2006.11371v2*.

Debnath, A. K., Compadre, R. L., Debnath, G., Shusterman, A. J., & Hansch, C. (1991). Structure-activity relationship of mutagenic aromatic and heteroaromatic nitro compounds. correlation with molecular orbital energies and hydrophobicity. *Journal of Medicinal Chemistry*, 34(2), 786–797.

Dieber, J., & Kirrane, S. (2020). *Why model why? Assessing the strengths and limitations of LIME*. Vienna, Austria: arXiv preprint arXiv 2012.

Dincer, A. B., Celik, S., Hiranuma, N., & Lee, S. I. (2018). DeepProfile: deep learning of cancer molecular profiles for precision medicine. *bioRxiv*, 278739

Ding, M. Q., Chen, L., Cooper, G. F., Young, J. D., & Lu, X. (2018). Precision oncology beyond targeted therapy: combining omics data with machine learning matches the majority of cancer cells to effective therapeutics. *Molecular Cancer Research*, 16(2), 269–278.

Feng, Q., Dueva, E., Cherkasov, A., & Ester, M. (2018). PADME: a deep learning-based framework for drug–target interaction prediction. *arXiv 2018*.

Ferdousi, R., Safdari, R., & Omidi., Y. (2016). Computational prediction of drug-drug interactions based on drugs functional similarities. *Journal of Biomedical Informatics*, 70, 54–64.

Gao, K. Y., Fokoue, A., Luo, H., Iyengar, A., Dey, S., & Zhang, P. (2017). Interpretable drug target prediction using deep neural representation. Proceedings of the International Joint Conference on Artificial Intelligence, 2018, 3371-3377. Melbourne Australia.

Ghorbani, A., Wexler, J., Zou, J. Y., & Kim, B. (2019). Towards automatic concept-based explanations. *Advances in Neural Information Processing Systems*, 32.

Gilpin, L. H., Bau, D., Yuan, B. Z., Bajwa, A., Specter, M., & Kagal, L. (2018). Explaining explanations: An overview of interpretability of machine learning. *IEEE 5th International Conference on Data Science and Advanced Analytics (DSAA)* (pp. 80–89). IEEE.

Gómez-Bombarelli, R., Wei, J. N., Duvenaud, D., Hernández-Lobato, J. M., Sánchez-Lengeling, B., Sheberla, D., … Aspuru-Guzik, A. (2018). Automatic chemical design using a data-driven continuous representation of molecules. *ACS Central Science*, 4(2), 268–276.

Gunning, D. (2017). *Explainable Artificial Intelligence (XAI)*. Defense Advanced Research Projects Agency (DARPA), 2(2), 1.

Hearst, M. A., Dumais, S. T., Osuna, E., Platt, J., & Scholkopf, B. (1998). Support vector machines. *IEEE Intelligent Systems and their applications*, 13(4), 18–28.

Hinton, G. E. (2012). A practical guide to training restricted Boltzmann machines. In *Neural Networks: Tricks of the Trade*, (pp. 599–619), Berlin, Heidelberg: Springer Berlin Heidelberg.

Hinton, G. E., Osindero, S., & Teh, Y. W. (2006). A fast learning algorithm for deep belief nets. *Neural Computation*, 18(7), 1527–1554.

Hochuli, J., Helbling, A., Skaist, T., Ragoza, M., & Koes, D. R. (2018). Visualizing convolutional neural network protein-ligand scoring. *Journal of Molecular Graphics and Modelling*, 84, 96–108.

Hsu, C.-C., Zhuang, Y.-X., & Lee, C.-Y. (2020). Deep fake image detection based on pairwise learning. *Applied Sciences*, 10(1), 370.

Ibrahim, M., Louie, M., Modarres, C., & Paisley, J. (2019). Global explanations of neural networks: Mapping the landscape of predictions. *Proceedings of the 2019 AAAI/ACM Conference on AI, Ethics, and Society*, Honolulu, HI, USA, (21 January, 2019) (pp. 279–287).

Jiménez-Luna, J., Grisoni, F., & Schneider, G. (2020). Drug discovery with explainable artificial intelligence. *Nature Machine Intelligence*, 2(10), 573–584.

Jordan, M. I., & Mitchell, T. M. (2015). Machine learning: trends, perspectives, and prospects. Science, 349(6245), 255–260.

Kamilaris, A., & Prenafeta-Boldú, F. X. (2018). Deep learning in agriculture: A survey. *Computers and Electronics in Agriculture*, 147, 70–90.

Kim, B., Wattenberg, M., Gilmer, J., Cai, C., Wexler, J., & Viegas, F. (2018). Interpretability beyond feature attribution: quantitative testing with concept activation vectors (tcav). *International Conference on Machine Learning* (pp. 2668–2677). PMLR.

LeCun, Y., Bengio, Y., & Hinton, G. (2015). Deep learning. Nature, 521(7553), 436–444.

Lee, I., Keum, J., & Nam, H. (2019). DeepConv-DTI: prediction of drug–target interactions via deep learning with convolution on protein sequences. *PLoS Computional Biology*, 15(6), 1–21.

Leelananda, S. P., & Lindert, S. (2016). Computational methods in drug discovery. *Beilstein Journal of Organic Chemistry*, 12(1), 2694–2718.

Lei, T., Barzilay, R., & Jaakkola, T. (2016). Rationalizing Neural Predictions. *Proceedings of the 2016 Conference on Empirical Methods in Natural Language Processing* (pp. 107–117). Austin, Texas: Association for Computational Linguistics.

Liu, S., Zhang, Y., Cui, Y., Qiu, Y., Deng, Y., Zhang, Z. M., & Zhang, W. (2022). Enhancing drug-drug interaction prediction using deep attention neural networks. *IEEE/ACM Transactions on Computational Biology and Bioinformatics*, 20(2), 976–985.

Lundberg, S. M., & Lee, S.-I. (2017). A unified approach to interpreting model predictions. In *Proceedings of the 31st Conference on Neural Information Processing Systems* (pp. 1–10), Long Beach, CA, USA, Dec. 4–9, 2017.

Luo, D., Cheng, W., Xu, D., Yu, W., Zong, B., Chen, H., & Zhang, X. (2020). Parameterized explainer for graph neural network. *Advances in Neural Information Processing Systems*, 33, 19620–19631.

Luo, Y., Zhao, X., Zhou, J., Yang, J., Zhang, Y., Kuang, W., … Zeng, J. (2017). A network integration approach for drug–target interaction prediction and computational drug repositioning from heterogeneous information. Nature Communications, 8(1), 573.

Mahendran, A., & Vedaldi, A. (2016). Salient deconvolutional networks. *Computer Vision–ECCV 2016: 14th European Conference* (pp. 120–135). Amsterdam, The Netherlands: Springer International Publishing.

Mahesh, B. (2020). Machine learning algorithms – a review. *International Journal of Science and Research (IJSR)*, 9(1), 381–386.

McCloskey, K., Taly, A., Monti, F., Brenner, M. P., & Colwell, L. J. (2019). Using attribution to decode binding mechanism in neural network models for chemistry. *Proceedings of the National Academy of Sciences*, 116(24), 11624–11629.

Mohseni, S., & Ragan, E. D. (2018). A human-grounded evaluation benchmark for local explanations of machine learning. *arXiv preprint arXiv:1801.05075*.

Mukhamediev, R. I., Symagulov, A., Kuchin, Y., Yakunin, K., & Yelis, M. (2011). From classical machine learning to deep neural networks: a simplified scientometric review. *Applied Sciences*, 11(12), 5541.

Nielsen, M. A. (2015). *Neural networks and deep learning.* San Francisco, CA: Determination Press.

Noh, H., Hong, S., & Han, B. (2015). Learning deconvolution network for semantic segmentation. *Proceedings of the IEEE International Conference on Computer Vision* (pp. 1520–1528). IEEE.

Phillips, P. J., Hahn, C. A., Fontana, P. C., Broniatowski, D. A., & Przybocki, M. A. (2020). *Four principles of explainable artificial intelligence.* Gaithersburg, Maryland: NIST, 18.

Pisner, D. A., & Schnyer, D. M. (2020). Support vector machine. In A. Mechelli, & S. Vieira (eds.), *Machine Learning* (pp. 101–121). Cambridge, MA, USA: Academic Press.

Pope, P. E., Kolouri, S., Rostami, M., Martin, C. E., & Hoffman, H. (2019). Explainability methods for graph convolutional neural networks. *Proceedings of the IEEE/CVF Conference on Computer Vision and Pattern Recognition* (pp. 10772–10781), Long Beach, CA, USA, June 16–20, 2019.

Quinlan, J. R. (1990). Decision trees and decision-making. *IEEE Transactions on Systems, Man, and Cybernetics*, 20(2), 339–346.

Robinson, A. J., & Voronkov, A. (2001). *Handbook of automated reasoning*. Amsterdam, Netherlands: Elsevier.

Rodríguez-Pérez, R., & Bajorath, J. (2019). Interpretation of compound activity predictions from complex machine learning models using local approximations and shapley values. *Journal of Medicinal Chemistry*, 63(16), 8761–8777.

Ryu, J. Y., Kim, H. U., & Lee, S. Y. (2018). Deep learning improves prediction of drug–drug and drug–food interactions. *Proceedings of the National Academy of Sciences,* 115(18), (pp. E4304–E4311). National Academy of Sciences.

Schlichtkrull, M. S., Cao, N. D., & Titov, I. (2020). Interpreting graph neural networks for NLP with differentiable edge masking. arXiv preprint *arXiv:2010.00577*.

Selvaraju, R. R., Cogswell, M., Das, A., Vedantam, R., Parikh, D., & Batra, D. (2017). Grad-cam: Visual explanations from deep networks via gradient-based localization. *Proceedings of the IEEE International Conference on Computer Vision* (pp. 618–626), Venice, Italy, October 22–29, 2017, IEEE.

Sharifi-Noghabi, H., Zolotareva, O., Collins, C. C., & Ester, M. (2019). MOLI: multi-omics late integration with deep neural networks for drug response prediction . *Bioinformatics*, 35(14), i501–i509.

Shrikumar, A., Greenside, P., & Kundaje, A. (2017). Learning important features through propagating activation differences. *Proceedings of the 34th International Conference on Machine Learning* (pp. 3145–3153). Sydney, NSW, Australia: JMLR.org.

Simonyan, K., Vedaldi, A., & Zisserman, A. (2014). Deep inside convolutional networks: visualising image classification models and saliency maps. In *Proceedings of the International Conference on Learning Representations (ICLR)*. ICLR (pp. 1–8). Banff, AB, Canada, April 14–16, 2014..

Sun, X., Ma, L., Du, X., Feng, J., & Dong, K. (2018). Deep convolution neural networks for drug–drug interaction extraction. *IEEE International Conference on Bioinformatics and Biomedicine (BIBM),* Madrid, Spain, 2018 (pp. 1662–1668), Madrid, Spain, December 3–6, 2018. IEEE.

Thafar, M., Raies, A. B., Albaradei, S., Essack, M., & Bajic, V. B. (2019). Comparison study of computational prediction tools for drug-target binding affinities. *Frontiers in Chemistry*, 7, 782.

Turri, V. (2022). What is Explainable AI? Retrieved 2023, from https://insights.sei.cmu.edu/blog/what-is-explainable-ai/

Wang, S., Zhou, T., & Bilmes, J. (2019). Bias also matters: Bias attribution for deep neural network explanation. *International Conference on Machine Learning* (pp. 6659–6667).

Wang, Y.-B., You, Z.-H., Yang, S., Yi, H.-C., Chen, Z.-H., & Zheng, K. (2020). A deep learning-based method for drug-target interaction prediction based on long short-term memory neural network. BMC Medical Informatics and Decision Making, 20(2), 1–9.

Wigmore, I. (2018). *Explainable AI (XAI)*. Retrieved 2023, from www.techtarget.com/whatis/definition/explainable-AI-XAI,

Wu, W., Su, Y., Chen, X., Zhao, S., King, I., Lyu, M. R., & Tai, Y.-W. (2020). Towards global explanations of convolutional neural networks with concept attribution. In *Proceedings of the IEEE/CVF Conference on Computer Vision and Pattern Recognition* (pp. 8652–8661), Los Alamitos, CA, USA, June 14–19, 2020.

Yamashita, R., Nishio, M., Do, R. K., & Togashi, K. (2018). Convolutional neural networks: an overview and application in radiology. *Insights into Imaging*, 9, 611–629.

Ying, Z., Bourgeois, D., You, J., Zitnik, M., & Leskovec, J. (2019). Gnnexplainer: Generating explanations for graph neural networks. In: NeurIPS 2019 Proceedings (Advances in Neural Information Processing Systems), (pp. 1–12), Vancouver, BC, Canada, December 8–14, 2019.

Zhai, J., Zhang, S., Chen, J., & He, Q. (2018). Autoencoder and its various variants. *IEEE International Conference on Systems, Man, and Cybernetics (SMC)*, Mayazaki, Japan,2018, (pp. 415–419), Mayazaki, Japan, October 7–10, 2018. IEEE.

Zhou, B., Khosla, A., Lapedriza, A., Oliva, A., & Torralba, A. (2016). Learning deep features for discriminative localization. *Proceedings of the IEEE Conference on Computer Vision and Pattern Recognition* (pp. 2921–2929), Las Vegas, NV, USA, June 26–July 1, 2016. IEEE.

Zia, T., & Zahid, U. (2019). Long short-term memory recurrent neural network architectures for Urdu acoustic modeling. *International Journal of Speech Technology*, 22, 21–30.

Zitnik, M., Agrawal, M., & Leskovec, J. (2018). Modeling polypharmacy side effects with graph convolutional networks. *Bioinformatics*, 34(13), i457–i466.

12 A Hybrid Explainable Artificial Intelligence Approach for Anti-Cancer Drug Discovery

Exploring the Potential of Explainable Artificial Intelligence in Computational Biology

K. Aditya Shastry

12.1 INTRODUCTION

Developing artificial intelligence (AI) systems that offer simple and comprehensible descriptions of their decision-making processes is the objective of explainable AI (XAI) (Zhang et al., 2022). Owing to the critical nature of interpreting information in computational biology, XAI is gaining ground (Krajna et al., 2022). Understanding the biotic behaviour at play, validating outcomes, and having faith in the estimates are all aided by AI algorithms that are easy to understand (Lötsch et al., 2022). This chapter's overarching goal is to assess the state of XAI in computational biology. This analysis will cover such topics as current methods, obstacles, and application studies.

In the domain of computational biology, XAI methods are utilized increasingly to aid in understanding complicated biological processes, making illness predictions, and creating novel therapies. Scientists may utilize XAI methods to better understand the key elements of massive physiological databases and the manner in which communication happens. In addition, XAI methods could be utilized to spot areas of improvement in AI techniques, letting scientists hone their techniques.

Here, we suggest a generic hybrid strategy to XAI-based malignancy medication discovery by fusing deep learning (DL) systems with causal inference methods. This method may help in the detection of novel therapeutic targets and focus on the interaction among the several elements that promote cancer growth. Additionally, this chapter offers a complete examination of state-of-the-art XAI methods in computational biology. We examine the usage of numerous XAI methods in computational

DOI: 10.1201/9781003426073-12

biology, including saliency maps, layer-wise relevance propagation, and decision trees. These methods could elucidate the principles of biology behind the information observed and explain how AI algorithms reach their judgments.

XAI approaches such as decision trees, linear models, and neural networks were all applied in computational biology. Because of their capacity to identify intricate connections and trends, DL methods are being extensively employed for evaluating biological information. Novel XAI approaches including saliency maps, decision trees, and layer-wise relevance propagation have been developed in reply to the challenges that using DL methods present with regards to clarity and interpretation.

There are many obstacles to using XAI in computational biology, such as the multifaceted nature of biological information and the necessity for comprehension in making choices, plus the necessity for openness during the procedure of analysis. Furthermore, the estimation of XAI systems lacks uniformity, making it hard to evaluate and compare various methodologies.

XAI demonstrates its usefulness in computational biology through several case studies, specifically in the examination of genetic information, proteomics information, along with metabolomic information. XAI is being used to anticipate the molecular makeup of proteins, which is vital to comprehending how they operate and to detect the most significant genetic material impacted in illnesses such as malignancy. Metabolomic information, which describes a cell's metabolism, is being analyzed with XAI to find diabetic indicators.

The most important results of this chapter are as follows:

- We propose a generic hybrid XAI approach to find new cancer treatments by combining DL models with inductive reasoning techniques.
- We perform a comprehensive analysis of state-of-the-art XAI methods in computational biology.
- We explore the use of XAI techniques in computational biology, such as saliency maps, layer-wise relevance transmission, and decision trees.
- We discuss the difficulties in understanding and trusting XAI models in computational biology.
- We show the possible solutions to the problems associated with the understanding and reliability of XAI systems.
- We prove the usefulness of XAI in computational biology by analyzing and discussing in depth numerous real-world scenarios.

The balance of the chapter is organized as follows. In Section 2, we'll take a look at the various XAI methods now in use in computational biology. In Section 3, we offer a generic hybridized approach that uses XAI to find novel cancer medicines by combining DL models with inference about causality techniques. In Section 4, we examine the difficulties and solutions of developing XAI simulations for computational biology. Section 5 details various practical applications of XAI in the field of computational biology. Section 6 provides a summary of potential future research topics. Section 7 serves as the conclusion to the work, summarizing its main points

12.2 XAI TECHNIQUES IN COMPUTATIONAL BIOLOGY

The following are examples of XAI methods employed in computational biology (Kim et al., 2021; Holzinger et al., 2022; Rout et al., 2022; Chaddad et al., 2023).

12.2.1 MODEL VISUALIZATION

When it comes to understanding how ML models work, model visualization is a crucial XAI approach utilized in computational biology. The following are some of the widely used types of visualization:

- Saliency maps: The most influential parts of an input image or feature on a model's predictions are highlighted in these heat maps. This may be useful for pinpointing the parts of an information set that hold the most predictive power.
- Activation maps: The firings of synapses in a network of neurons can be visualized using activation maps, which reveal what elements of an input image or attribute are going through processing by the network.
- Layer-wise relevance propagation (LRP): LRP is a method for visualizing the contributions of individual neurons and features to the forecasts of a model (Montavon et al., 2019). It involves propagating the relevance of a prediction back through the layers of a neural network to regulate which inputs are most important for making a prediction.

These visualization techniques deliver useful insights on the performance and decisions of computational biology models, and could assist in improving the interpretability and reliability of these models. Additionally, they can support scientific discovery by giving a greater awareness of the interactions and patterns in biological data, and by identifying the key factors that contribute to specific predictions (Cho et al., 2020).

12.2.2 FEATURE IMPORTANCE ANALYSIS

Feature importance analysis is a commonly used XAI technique in computational biology for determining the relative importance of different features or variables in a dataset (Zacharias et al., 2022). There are various approaches for performing feature importance analysis including the following:

- Permutation feature importance: This method involves randomly transposing the values of a single feature in the dataset and observing the effect on the model's performance. The feature that has the highest effect on the model's operation is considered the most important (Muschalik et al., 2022).
- Lasso and ridge regression: These regularization procedures are employed to assign importance scores to individual features by penalizing their coefficients, with larger penalties assigned to less important features (Sai et al., 2022).
- Random forest feature importance: Random forest is a ML method that could be utilized to measure the importance of different features in the dataset. It

does this by computing the average reduction in impurity for each feature, with the most important features having the greatest reduction (Lukyanenko et al., 2020).

*   SHapley Additive exPlanations (SHAP): It is a method for computing feature importance that considers the interactions across different attributes in the dataset (Abeyagunasekera et al., 2022).

By determining the comparative significance of distinct attributes in a dataset, feature importance analysis can offer useful perceptions into the biological processes that drive the predictions of computational biology models. It can also support the development of more accurate and reliable models by identifying the most important features to include in the dataset.

12.2.3 COUNTERFACTUAL ANALYSIS

Counterfactual analysis is a XAI technique that involves generating alternate scenarios based on the forecasts of a ML technique and examining the effects of changing specific input parameters on the outcome. In computational biology, this method may be used to acquire greater insight of the relationships between different biological attributes and the model's predictions, and to identify the key factors that contribute to a specific prediction (Thiagarajan et al., 2022).

For example, during the prediction of the efficacy of a drug, a counterfactual analysis might involve examining the effects of changing the dose or administration schedule of the drug on the model's predictions. This may offer perceptions into the most effective dosing regimen and the fundamental biotic instruments that lead to the predictions (Mertes et al., 2022).

Counterfactual analysis may be employed to test the robustness and generalizability of computational biology models, and to identify potential sources of bias or error in the model's predictions. Additionally, it could offer a more transparent and interpretable justification for the model's behavior, which can help to build belief and assurance in the model among stakeholders (Chou et al., 2022).

12.2.4 MODEL INTERPRETABILITY ALGORITHMS

There exist numerous model interpretability processes that are regularly utilized in computational biology to offer insights into the workings of ML models (Farahani et al., 2022):

*   Local Interpretable Model-Agnostic Explanations (LIME): It generates a justification for the estimates of any ML model by fitting a simple linear model to the neighborhood of the prediction (Shi et al., 2022).
*   Partial Dependence Plots (PDP): PDPs are graphical representations of the relationship between a feature and a model's prediction, possessing other attributes held persistent. They provide a visual representation of the influence of distinct traits on the prediction (Zhang et al., 2018).

- Decision Trees: These are a straightforward and intuitive form of interpretable ML that are employed to provide a transparent explanation for the decisions of a model (Gerlach et al., 2022).

These algorithms can help to deliver greater insight of the workings of computational biology models and can support scientific discovery by identifying the key factors that contribute to specific predictions. Additionally, they aid in enhancing the interpretability and reliability of these models by identifying potential sources of bias or error in the predictions.

12.2.5 EXPLAINABLE DEEP LEARNING

Explainable deep learning in computational biology refers to the usage of XAI techniques to provide insights into the workings of DL models applied to biological data (Yang et al., 2022). Deep learning is an AI-based subfield of machine learning (ML) used to understand intricate associations across inputs and outputs. While DL algorithms have accomplished noteworthy feats in a diversity of applications, their behavior can often be difficult to understand, especially for complex representations possessing several hidden layers (Liu et al., 2022).

To address this issue, XAI methods are developed to generate justifications for the predictions of DL models in computational biology. These methods can be utilized to detect the key attributes in the input information that contribute to an explicit calculation, and to visualize the workings of the model at different levels of abstraction, such as individual neurons and layers (Watson, 2022).

For example, layer-wise relevance propagation (LRP) is a commonly used XAI technique for explaining DL models. LRP could be utilized to detect the regions of an image that are most important for a prediction, or to determine the specific proteins that are most important for a specific biological prediction (Ullah et al., 2022).

Explainable DL can help to improve the interpretability and transparency of DL models in computational biology, and can support scientific discovery by providing insights into the biological processes that drive the predictions of these models. Additionally, it can help to build trust and confidence in the models by providing a clear and understandable explanation for their behavior (Ras et al., 2022).

Each of these techniques has its own strengths and limitations, and the choice of XAI technique to use will depend on the specific requirements of the computational biology problem being addressed.

12.3 GENERIC HYBRID APPROACH FOR DISCOVERING ANTI-CANCER DRUGS USING XAI

In this section, we suggest a novel approach for discovering anti-cancer drugs using XAI. One possible novel approach for discovering anti-cancer drugs using XAI could be a combination of DL models and causal inference algorithms. Discovering anti-cancer drugs is a complex and multi-step process that requires collaboration between multiple disciplines such as pharmacology, cancer biology, and computer science.

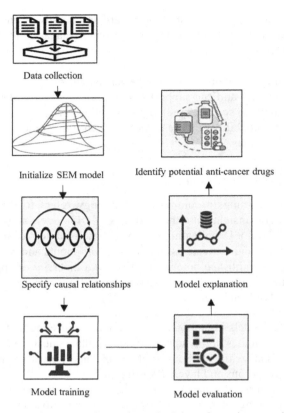

FIGURE 12.1   Proposed XAI-based generic methodology for anti-cancer drug discovery.

Figure 12.1 shows the proposed generic hybrid approach for discovering anti-cancer drugs using XAI.

Here are the detailed steps for a novel approach for discovering anti-cancer drugs using XAI:

12.3.1   DATA COLLECTION

The first step in the process is to collect a large dataset of molecular information relevant to cancer biology and pharmacology. This data can come from a variety of resources like genomic databases, transcriptomic data, proteomic data, and clinical trial records. The data should be preprocessed to ensure that it is of high quality and ready for analysis.

12.3.2   INITIALIZE STRUCTURAL EQUATION MODEL

The initialize structural equation model (SEM) step is an important step in the development of a novel XAI approach for drug discovery in blood cancer. The aim of this step is to build a mathematical model that represents the relationships between molecular data, drug efficacy, and the underlying mechanisms of disease.

Mathematically, a structural equation model (SEM) can be represented as a directed acyclic graph (DAG) where nodes represent variables and edges represent relationships between variables. In the context of drug discovery for blood cancer, the SEM could represent variables such as molecular data (e.g., gene expression, epigenetic marks, etc.), drug efficacy (e.g., response rate, progression-free survival, etc.), and potential underlying mechanisms of disease (e.g., pathways, signaling pathways, etc.).

To initialize the SEM, the following steps can be followed:

- Define the variables: The first step is to define the variables that will be included in the SEM. This includes molecular data, drug efficacy, and the underlying mechanisms of disease.
- Specify relationships between variables: Once the variables have been defined, the relationships between them can be specified. For example, the molecular data may influence drug efficacy, and the underlying mechanisms of disease may influence both molecular data and drug efficacy.
- Establish measurement models: For each variable in the SEM, a measurement model must be established. This involves specifying how the variable will be observed and measured in the data.
- Specify structural models: The final step in initializing the SEM is to specify the structural models, which represent the relationships between variables in the SEM. This can be done using regression models, path analysis, or other statistical methods.

Algorithm 1 provides a general outline for initializing the SEM:

Algorithm 1: *initialize-sem(molecular-data, drug-efficacy):*
{ # Define variables
   variables = ['molecular-data', 'drug-efficacy', 'underlying-mechanisms']
   # Specify relationships between variables
   relationships = [('molecular-data', 'drug-efficacy'),
        ('underlying-mechanisms', 'molecular-data'),
        ('underlying-mechanisms', 'drug-efficacy')]
      # Establish measurement models
Measurement-models = {'molecular-data': linear-regression, 'drug-efficacy': logistic_regression, 'underlying-mechanisms': pathway-analysis}
   # Specify structural models
structural-models = {('molecular-data', 'drug-efficacy'): multiple-regression,
        ('underlying-mechanisms', 'molecular-data'): pathway-analysis,
        ('underlying-mechanisms', 'drug-efficacy'): multiple-logistic-regression}
   # Initialize SEM
sem = StructuralEquationModel(variables, relationships, measurement-models, structural-models)
return sem }

The initialize structural equation model (SEM) step involves setting up a statistical model that can represent the relationships between variables in the data. In the context of drug discovery in cancer, the variables may include molecular data (e.g., gene expression, mutation status) and drug efficacy data (e.g., response to a specific drug).

The SEM framework allows us to specify the relationships between variables using a series of equations. Each equation represents a linear regression model, where the dependent feature is linked to one or more autonomous features. Equation (12.1) demonstrates this concept:

$$P = C_0 + C_1 I_1 + C_2 I_2 + \ldots + C_n I_n + \text{Err} \tag{12.1}$$

where P is the dependent feature, I_1, I_2, ..., I_n are the autonomous features, C_0, C_1, C_2, ..., C_n are the coefficients, and Err is the error term. These equations can be combined to form a larger model that represents the relationships between multiple variables.

The SEM framework allows us to estimate the coefficients in these equations and use them to make predictions about the relationships between variables. This can provide valuable insights into the causal relationships between molecular data and drug efficacy, which can inform the discovery of new anti-cancer drugs.

12.3.3 SPECIFY CAUSAL RELATIONSHIPS

This step in the context of discovering anti-cancer drugs using XAI involves determining the relationships between the various molecular variables and drug efficacy in the SEM. This can be accomplished through a combination of statistical and causal inference techniques, including Bayesian networks and causal effect models.

To specify the causal relationships in the SEM, we need to estimate the causal effect of each molecular variable on drug efficacy. This can be done using a variety of methods, including regression analysis, counterfactual inference, and causal inference algorithms such as the g-formula or the inverse probability weighting (IPW) method.

Once the causal effects of each molecular variable on drug efficacy have been estimated, we can use this information to construct a directed acyclic graph (DAG) that represents the causal relations in the SEM. The DAG can be used to identify the most important variables that influence drug efficacy and to determine the causal pathways through which they exert their effects.

The algorithm for specifying causal relationships in the SEM is represented in algorithm 2.

Algorithm 2: specify-causal-relationships(molecular-data, drug-efficacy):
{ # Estimate causal effect of each molecular variable on drug efficacy
  for variable in molecular-data:
  estimate-causal-effect(variable, drug-efficacy)
 # Construct directed acyclic graph (DAG) to represent causal relationships in SEM

DAG = construct-DAG(molecular-data, drug-efficacy)
 # Identify most important variables that influence drug efficacy
Important-variables = identify-important-variables(DAG)
 # Determine causal pathways through which important variables exert their effects
causal-pathways = determine-causal-pathways(important-variables, DAG)
return important-variables, causal-pathways }

12.3.4 TRAIN THE MODEL

The subsequent phase is to train the SEM utilizing the molecular data and drug efficacy information. This involves fitting the SEM to the data to find the best parameters that represent the relationships between the variables. Algorithm 3 for the train the model step for discovery of anti-cancer drugs using XAI is shown below.

Algorithm 3: train-model(SEM, molecular-data, drug-efficacy):
{ # Fit the SEM to the molecular_data and drug_efficacy data using maximum likelihood estimation
SEM.fit(molecular-data, drug-efficacy)
 # Compute the goodness of fit measures for the SEM
gof-measures = SEM.compute-gof-measures()
 # Evaluate the performance of the SEM on a validation set
performance = SEM.evaluate-on-validation-set()
 # If the performance of the SEM is not satisfactory, modify the SEM and repeat the training process
while performance is not satisfactory:
SEM.modify()
SEM.fit(molecular-data, drug-efficacy)
gof-measures = SEM.compute-gof-measures()
performance = SEM.evaluate-on-validation-set()
#Return the trained SEM
return SEM }

In algorithm 3, SEM is the initialized SEM, molecular data is the input data for molecular features, and drug efficacy is the target data for drug efficacy. The fit function is used to fit the SEM to the data using maximum likelihood estimation. The compute-gof-measures function is used to compute the goodness-of-fit measures for the SEM, and the evaluate on the validation-set function is used to assess the functioning of the SEM on a validation set. If the performance of the SEM is not satisfactory, the SEM is modified and the training procedure is repeated. Finally, the trained SEM is returned.

12.3.5 EVALUATE MODEL PERFORMANCE

The performance of the SEM is evaluated to determine the accuracy of the model in predicting drug efficacy. This is done by comparing the predictions made by the model with the actual drug efficacy data. Algorithm 4 shows the XAI model evaluation.

> *Algorithm 4: evaluate-model-performance(model, molecular-data,*
> *drug-efficacy):*
> *{ # Divide the molecular information into train and test datasets*
> *train-data, test-data = split-data(molecular-data)*
> *# Split the drug efficacy data into train and test datasets*
> *train-target, test-target = split-data(drug-efficacy)*
> *# Train the model utilizing the training datasets*
> *model.fit(train-data, train-target)*
> *# Use the trained model to make forecasts on the test datasets*
> *predictions = model.predict(test-data)*
> *# Assess the model's performance by calculating the mean squared*
> *   error (MSE)*
> *mse = MSE(test-target, predictions)*
> *   print("Mean Squared Error:", mse)*
> *return mse }*

The purpose of algorithm 4 is to evaluate the performance of a given model on a dataset consisting of molecular information and drug efficacy data. The algorithm takes as input the trained model, molecular data, and drug-efficacy data. The algorithm first splits the molecular data into train and test datasets, and also splits the drug efficacy data into train and test datasets.

The algorithm then trains the model on the training datasets using the fit method. Once the model is trained, it makes forecasts on the test datasets using the predict method. The algorithm then assesses the performance of the model by calculating the mean squared error (MSE) between the predicted drug efficacy values and the actual drug efficacy values in the test dataset.

The algorithm prints the MSE value and returns it as the output of the algorithm. The MSE value indicates how well the model performs on the test dataset, with lower MSE values indicating better performance. This algorithm can be used to compare the performance of different models on the same dataset or to evaluate the performance of a single model on different dataset.

12.3.6 EXPLAIN MODEL PREDICTIONS

The final step is to explain the predictions made by the model. This involves using XAI methods to provide insights into why the model made certain predictions and how the molecular information influenced the drug efficacy. This step is important for ensuring that the model is trustworthy and provides meaningful insights into the drug discovery process. Algorithm 5 shows the explanatory nature of the XAI model.

Algorithm 5 explain-model-predictions(model, molecular-data, drug-efficacy):

{ # Input: Trained model, molecular data and drug efficacy data
 # Output: Explanation of the model predictions
 # Predict the efficacy of drugs on molecular data
drug-predictions = model.predict(molecular-data)
 # Use the XAI technique to explain the predictions
explanations = XAI-technique(model, molecular-data, drug-predictions)
 # Compare the explanations with the actual drug efficacy data
evaluation = compare-explanations-with-actual-data(explanations,
*   drug-efficacy)*
 # Return the evaluation results
return evaluation }

12.3.7 IDENTIFY POTENTIAL ANTI-CANCER DRUGS

Once the SEM has been trained and evaluated, it can be used to identify potential anti-cancer drugs. This involves using the model to predict the efficacy of new compounds and selecting the compounds with the highest predicted efficacy for further investigation. Algorithm 6 shows the identification of potential anti-cancer drugs.

Algorithm 6: *identify-potential-anti-cancer-drugs(model, molecular-data, drug-efficacy):*
{ predictions = predict(model, molecular-data)
confidence-scores = calculate-confidence-scores(model, predictions)
potential-drugs = []
for i in range(len(molecular-data)):
if confidence-scores[i] > threshold and drug-efficacy[i] == "effective":
potential-drugs.append (molecular-data[i])
return potential-drugs }

In algorithm 6, the identify-potential-anti-cancer-drugs function takes in a trained model, molecular data, and drug-efficacy data as inputs. The function starts by using the predict function to get predictions for each molecular data point. Then, it uses the calculate-confidence-scores function to get the confidence scores for each prediction. Next, the function loops over each molecular data point and checks if its confidence score is above a specified threshold and its drug efficacy label is effective. If these conditions are met, the molecular data is added to the potential-drugs list. Finally, the function returns the list of potential anti-cancer drugs.

This novel approach for discovering anti-cancer drugs using XAI can be an effective and efficient way of identifying new potential drugs for the treatment of blood cancer and other types of cancer. By combining the strengths of AI and XAI, this approach provides a comprehensive and transparent solution for the drug discovery process.

12.4 CHALLENGES ASSOCIATED WITH THE INTERPRETATION AND TRUSTWORTHINESS OF XAI MODELS AND POSSIBLE STRATEGIES

The usage of XAI models in computational biology poses several challenges that can affect the interpretation and trustworthiness of these models. Some of these challenges are (Han & Liu, 2021; Sapoval et al., 2022):

- Lack of transparency: XAI models in computational biology can be complicated and difficult to interpret, making it difficult to realize how they make predictions and decisions.
- Bias in data and models: The data used to train XAI models in computational biology can be biased, leading to incorrect or incomplete predictions. The models themselves may also have built-in biases that affect their predictions.
- Overfitting: XAI models in computational biology can overfit to the training data, leading to poor performance when applied to new data.
- Lack of robustness: XAI models in computational biology can be sensitive to small changes in the input data, leading to unreliable predictions.

Possible strategies to overcome these challenges include:

- Enhancing transparency: Developing XAI models that are transparent and easy to interpret can help improve the trustworthiness of these models.
- Addressing bias: Using diverse and representative datasets to train XAI models and incorporating fairness constraints in the model training process can help mitigate the impact of bias.
- Preventing overfitting: Regularization techniques, such as early stopping or dropout, can be used to prevent overfitting and improve the generalization performance of XAI models.
- Improving robustness: Evaluating XAI models on multiple datasets and under different conditions can help improve their robustness and reliability.

Table 12.1 illustrates the outline of these challenges and possible solutions.

Table 12.1 provides a summary of some of the challenges and strategies, but it is not exhaustive, and the best solution may vary depending on the specific use case. Overall, addressing these challenges requires a multi-disciplinary approach, involving experts in computational biology, machine learning, and ethics.

12.5 REAL-WORLD CASE STUDIES OF XAI IN COMPUTATIONAL BIOLOGY

12.5.1 CASE STUDY-1: XAI IN DRUG DISCOVERY

XAI has been applied in various areas of drug detection, including target detection, lead discovery, and drug repurposing. XAI helps researchers and practitioners better understand the decision-making processes of ML models, which is important in drug

TABLE 12.1

Challenges and Strategies for Building XAI Models in Computational Biology

Challenges	Possible Strategies
Lack of transparency and interpretability	Implement techniques such as Local Interpretable Model-Agnostic Explanations (LIME) and Gradient-weighted Class Activation Mapping (Grad-CAM) to offer graphic descriptions of model projections.
Bias in training data	Ensure the representativeness and diversity of the training dataset, and use bias correction techniques
Overconfidence in predictions	Use ambiguity quantification techniques such as Monte Carlo dropout to estimate the model's confidence in its predictions.
Lack of generalization to new data	Regularize the model to prevent overfitting, and assess the model's performance on a held-out test set to ensure its generalization ability.
Limited domain expertise	Collaborate with domain experts to interpret the results and validate the predictions of the model.
Difficulty in verifying results	Use multiple validation techniques and benchmark the model against existing methods.

discovery where decisions can have a major impact on human health. Here is a real-world case study that demonstrates the use of XAI in drug discovery:

- XAI-Assisted Drug Discovery: In Jiménez-Luna et al. (2020), the researchers used XAI to enhance the discovery of new drugs. The study used a deep neural network trained on a large dataset of molecular structures and biological activities to calculate the biologic movement of novel combinations. The XAI techniques were used to understand how the model made its predictions, which helped the researchers to gain insights into the underlying biological mechanisms and prioritize compounds for further testing.
- XAI-assisted Drug Repositioning: In Masuda & Mimori (2022), the researchers used XAI algorithms to assist in drug repositioning, where existing drugs are repurposed for new indications. The XAI model analyzed large datasets of genetic, biochemical, and pharmacological information to identify potential new uses for existing drugs. The results showed that the XAI model outperformed traditional methods in identifying new drug indications and led to the finding of new therapeutic opportunities
- XAI-based Virtual Screening of Drug Candidates: In Kimber et al., (2021) XAI algorithms were used to perform virtual screening of drug candidates, which is a computational method used to identify potential drug targets. The XAI model analyzed large datasets of molecular structures and identified potential new drug candidates. The results showed that the XAI model outperformed

traditional methods in identifying drug targets and headed to the detection of different medicine candidates with improved efficacy and safety.

- XAI-assisted Discovery of Adverse Drug Reactions: In Létinier et al. (2021), XAI algorithms were used to identify adverse drug reactions (ADRs) that were previously unknown. The XAI model analyzed large datasets of electronic health records and identified ADRs that were not previously reported in clinical trials. Their findings revealed that the XAI model could identify ADRs that were missed by traditional methods and improved patient safety by allowing for earlier detection and prevention of adverse reactions.
- XAI-based Predictive Modeling of Drug-Drug Interactions: In Vo et al. (2022), XAI algorithms were used to develop predictive models of drug-drug interactions, which are interactions between two or more drugs that can lead to adverse reactions. The XAI model analyzed large datasets of drug-drug interactions and identified potential interactions that were previously unknown. The results showed that the XAI model outperformed traditional methods in identifying drug-drug interactions and improved patient safety by allowing for earlier detection and prevention of adverse reactions.

These studies demonstrate the potential for XAI to play a major role in discovering medicines and their progress, by improving the accuracy and efficiency of drug discovery processes, and by increasing patient safety by identifying adverse drug reactions and drug-drug interactions.

12.5.2   Case Study-2: XAI in Predictive Modeling of Gene Regulation

XAI has been applied in various areas of gene regulation, including gene expression analysis, gene network inference, and gene function prediction. XAI helps researchers and practitioners better understand the decision-making processes of ML techniques, which is important in gene regulation where decisions can have a major impact on human health. Following are some real-world case studies related to XAI in predictive modeling of gene regulation:

- Identification of transcription factor binding sites: XAI has been used to develop predictive models that can identify the specific regions of DNA where transcription factors bind. This information is crucial for understanding gene regulation and can be used to predict how changes in transcription factor binding will impact gene expression. In Cao et al. (2022), XAI was applied to develop a predictive model that could detect the binding sites of the p53 transcription component, which performs a vital part in controlling the cell sequence and responding to DNA damage.
- Predicting gene expression patterns: XAI has been used to develop predictive models that can predict gene expression patterns based on genomic data. In Zhao et al. (2022), XAI was used to develop a predictive model that could identify the gene expression patterns associated with specific diseases, such as cancer. The XAI component of the model allowed researchers to understand the

genomic factors that were contributing to the gene expression patterns, which could be used to inform the development of targeted treatments for the disease.

- Analysis of non-coding RNA: XAI has been used to analyze the functionality of non-coding RNA in gene control. Non-coding RNA molecules are essential for controlling the transcription of genes, but their mechanisms of action are not well understood. XAI has been used to develop predictive models that can identify the specific regulation of expression of genes by compounds that lack a sequence, and how they are impacting the expression of specific genes (Lin & Wichadakul, 2022).

12.5.3 CASE STUDY-3: XAI IN PERSONALIZED MEDICINE

XAI in personalized medicine refers to the use of AI algorithms that can provide insights into their predictions or decisions, making the results of their analysis more transparent and interpretable to physicians and patients. This approach is especially important in the field of personalized medicine, where the goal is to provide individualized care tailored to a patient's unique needs, rather than relying on generalizations from large population datasets. Some case studies are as follows:

- Prediction of treatment response in cancer patients: Researchers used XAI to develop a ML algorithm that could predict how cancer patients would respond to different treatments based on their genomic data (Chakraborty et al., 2021). The algorithm was able to provide more accurate predictions than traditional approaches, and the XAI component allowed physicians to understand the reasons behind the predictions, leading to more informed treatment decisions.
- Risk assessment for cardiovascular disease: The study by Guleria et al. (2022) used XAI to analyze electronic health records (EHRs) and demographic data to predict the threat of cardiovascular disease in patients. The XAI algorithm was able to provide more accurate risk assessments than traditional approaches, and the XAI component allowed physicians to understand the factors contributing to a patient's risk, such as age, blood pressure, and cholesterol levels.
- Identifying genetic causes of rare diseases: XAI has been used to analyze genomic data to detect the inherited origins of uncommon disorders. In Brasil et al. (2019), a family with a rare genetic disorder was able to use XAI to identify the specific genetic mutations responsible for the disease in their family. This information helped physicians develop a more targeted and effective treatment plan for the affected individuals.
- Personalized treatment for depression: In Joyce et al. (2023), the authors used XAI to analyze brain imaging data and demographic information to predict how patients with depression would respond to different treatments. The XAI algorithm was able to provide more accurate predictions than traditional approaches, and the XAI component allowed physicians to understand the underlying mechanisms contributing to a patient's depression and how different treatments may impact their brain function.

12.5.4 CASE STUDY-4: XAI IN THE PREDICTION OF THE STRUCTURE OF PROTEINS

XAI could be applied to the field of protein structure prediction to provide more accurate and interpretable predictions. Forecasting the composition of proteins is the process of determining the three-dimensional structure of a protein determined by the order of its constituent amino acids. This information is crucial for understanding the function of proteins and can be used to inform the design of new proteins with specific functions. Some case studies on the use of XAI in protein structure prediction are summarized as follows:

- Improving accuracy of protein structure prediction: XAI has been used to develop ML methods to predict the 3D structure of proteins with greater accuracy. In one case, XAI was used to develop a predictive model that could accurately foresee the protein structure built on their amino acid sequences. The XAI component of the model allowed researchers to identify the interactions among the amino acid sequence and the protein structure, which could be used to inform the design of new proteins with specific functions (Pakhrin et al., 2021).
- Understanding the relationships between sequence and structure: XAI has been used to provide insights into the relationships between protein sequence and structure. In Anguita et al. (2020), XAI was used to analyze the structural changes that occur in proteins when their amino acid sequences are mutated. The XAI component of the analysis allowed researchers to understand the specific amino acid residues that are contributing to the structural changes, and how they may impact the function of the protein.
- Predicting protein-protein interactions: XAI has been used to develop predictive models that can identify the specific protein-protein interactions that are involved in cellular processes. In Madan et al. (2022), XAI was used to develop a predictive model that could identify the interactions between specific proteins and their targets, which could be used to understand the mechanisms of cellular processes and inform the development of targeted treatments for diseases.

12.5.5 CASE STUDY-5: XAI IN PREDICTING PROTEIN-PROTEIN INTERACTIONS UTILIZING GRAPH NEURAL NETWORKS (GNNS)

XAI can be applied to the likelihood of protein-protein interactions using GNNs. The protein-protein linkages are vital to a wide variety of biological functions, including signaling pathways, metabolic pathways, and the regulation of gene expression. Accurately predicting protein-protein communications is crucial for knowing the methods of these processes and can inform the development of targeted treatments for diseases. GNNs are a type of ML model that are well-suited for predicting protein-protein interactions, as they can handle the complex and dynamic relationships between proteins. By incorporating XAI into these models, researchers can better understand the relationships between the proteins that are being predicted to interact and the specific features that are contributing to the predictions.

There have been several real-world case studies that define the potential of XAI in predicting protein-protein interactions utilizing GNNs. Here are a few examples:

- Predicting protein-protein interactions in yeast: In Li et al. (2022), XAI was used to develop a GNN-based model for predicting protein-protein interactions in yeast. The XAI component of the model allowed researchers to understand the specific features of the yeast proteins that were contributing to the predictions, such as amino acid sequences, functional domains, and evolutionary information. The predictions made by the model were found to be more accurate than those made by traditional models, and the XAI component provided insights that could be used to improve the accuracy of future predictions.
- Improving the accuracy of protein-protein interaction predictions: In Zhou et al. (2022), XAI was used to develop a GNN-based model for forecasting protein-protein communications in human cells. The XAI component of the model allowed researchers to understand the specific features of the human proteins that were contributing to the predictions, such as amino acid sequences, functional domains, and evolutionary information. The predictions made by the model were found to be more accurate than those made by traditional models, and the XAI component provided insights that could be used to improve the accuracy of future predictions.
- Predicting protein-protein interactions in cancer: In Philipp et al. (2023), XAI was used to develop a GNN-based model for predicting protein-protein interactions in cancer cells. The XAI component of the model allowed researchers to understand the specific features of the cancer proteins that were contributing to the predictions, such as amino acid sequences, functional domains, and evolutionary information. The predictions made by the model were found to be more accurate than those made by traditional models, and the XAI component provided insights that could be used to improve the accuracy of future predictions and inform the development of targeted treatments for cancer.

12.6 FUTURE RESEARCH DIRECTIONS OF XAI IN COMPUTATIONAL BIOLOGY

XAI possesses the capacity to significantly advance the field of computational biology by providing more accurate and interpretable predictions and insights. There are several promising research directions in XAI for computational biology that have the potential to drive future developments in the field (Saeed & Omlin, 2023; Belle & Papantonis, 2021):

- Improving the accuracy of predictions: One of the primary goals of XAI in computational biology is to enhance the precision of predictions made by ML algorithms. This can be achieved by incorporating XAI techniques into existing methods to learn more about the specific features of the data that are contributing to the predictions, and by developing new models that incorporate XAI from the ground up.

- Interpreting complex biological systems: Another important goal of XAI in computational biology is to help researchers better understand complex biological systems. This can be achieved by developing XAI models that are capable of modeling complex relationships between proteins, genes, and other biological entities.
- Designing new therapies: XAI possesses the ability to considerably impact the development of new therapies for diseases by providing more accurate predictions and insights into the relationships between proteins, genes, and other biological entities. XAI models can be used to inform the design of new therapies by identifying the specific proteins, genes, and other biological entities that are contributing to disease states, and by providing insights into the mechanisms of these diseases.
- Incorporating multi-omics data: Another promising research direction in XAI for computational biology is the integration of multiple sources of data, such as genomics, transcriptomics, and proteomics data, to make predictions and gain insights into complex biological systems. This can be achieved by developing XAI models that are capable of integrating multiple sources of data and by incorporating XAI techniques into these models to provide interpretable predictions and insights.

Bridging the gap between AI and biology: A key challenge in XAI for computational biology is bridging the gap between AI and biology. This can be achieved by developing XAI models that are more accessible and interpretable to researchers in the biological sciences, and by incorporating XAI into existing workflows and pipelines used in computational biology.

These are just a few of the many exciting research directions in XAI for computational biology, and there is much more work to be done in this field. With continued advancements in XAI and computational biology, there is the potential to make significant impacts in the discovery of new treatments for diseases and in our understanding of complex biological systems.

12.7 CONCLUSION

In conclusion, the development of XAI is increasingly becoming important in computational biology as the interpretation of results is crucial in this field. XAI techniques can help researchers gain insights into complex biological systems, predict disease outcomes, and develop new treatments. This comprehensive research assessed the progress of XAI in computational biology, including the techniques used, the challenges faced, and the case studies that demonstrate its utility. The research proposed a generic hybrid approach for discovering anti-cancer drugs using XAI by integrating DL models and causal inference algorithms. Furthermore, the research presented an exhaustive review of advanced XAI techniques for computational biology, highlighting the challenges associated with the interpretation and trustworthiness of XAI models in this field. The case studies demonstrated the effectiveness of XAI in computational biology, which can provide insights into the underlying biological processes, help in the validation of results, and increase the trustworthiness of the models. The findings of this chapter have significant implications for the development of AI algorithms in computational biology and can contribute to the development of more accurate and robust models.

REFERENCES

Abeyagunasekera, S. H. P., Perera, Y., Chamara, K., Kaushalya, U., Sumathipala, P., & Senaweera, O. (2022). *LISA: Enhance the Explainability of Medical Images Unifying Current XAI Techniques.* In 2022 IEEE 7th International conference for Convergence in Technology (I2CT) (pp. 1–9). Mumbai, India. doi: 10.1109/I2CT54291.2022.9824840.

Anguita-Ruiz, A., Segura-Delgado, A., Alcalá, R., Aguilera, C. M., & Alcalá-Fdez, J. (2020). eXplainable Artificial Intelligence (XAI) for the identification of biologically relevant gene expression patterns in longitudinal human studies, insights from obesity research. *PLOS Computational Biology,* 16(4), e1007792. https://doi.org/10.1371/journal.pcbi.1007792.

Belle, V., & Papantonis, I. (2021). Principles and practice of explainable machine learning. *Frontiers in Big Data,* 4, 688969. https://doi.org/10.3389/fdata.2021.688969.

Brasil, S., Pascoal, C., Francisco, R., Dos Reis Ferreira, V., Videira, P. A., & Valadão, A. G. (2019). Artificial Intelligence (AI) in rare diseases: Is the future brighter? *Genes (Basel),* 10(12), 978. https://doi.org/10.3390/genes10120978.

Cao, L., Liu, P., Chen, J., & Deng, L. (2022). Prediction of Transcription factor binding sites using a combined deep learning approach. *Frontiers in Oncology,* 12, 893520. https://doi.org/10.3389/fonc.2022.893520.

Chaddad, A., Peng, J., Xu, J., & Bouridane, A. (2023). Survey of explainable AI techniques in healthcare. *Sensors,* 23(2), 634. https://doi.org/10.3390/s23020634.

Chakraborty, D., Ivan, C., Amero, P., Khan, M., Rodriguez-Aguayo, C., Başağaoğlu, H., & Lopez-Berestein, G. (2021). Explainable artificial intelligence reveals novel insight into tumor microenvironment conditions linked with better prognosis in patients with breast cancer. *Cancers,* 13(14), 3450. https://doi.org/10.3390/cancers13143450.

Cho, H., Lee, E. K., & Choi, I. S. (2020). Layer-wise relevance propagation of InteractionNet explains protein–ligand interactions at the atom level. *Scientific Reports,* 10, 21155. https://doi.org/10.1038/s41598-020-78169-6.

Chou, Y.-L., Moreira, C., Bruza, P., Ouyang, C., & Jorge, J. (2022). Counterfactuals and causability in explainable artificial intelligence: Theory, algorithms, and applications. *Information Fusion,* 81, 59–83. https://doi.org/10.1016/j.inffus.2021.11.003.

Farahani, F. V., Fiok, K., Lahijanian, B., Karwowski, W., & Douglas, P. K. (2022). Explainable AI: A review of applications to neuroimaging data. *Front Neuroscience,* 16, 906290. doi: 10.3389/fnins.2022.906290.

Gerlach, J., Hoppe, P., Jagels, S., Licker, L., & Michael H. B. (2022). Decision support for efficient XAI services – A morphological analysis, business model archetypes, and a decision tree. *Electron Markets,* 32, 2139–2158. https://doi.org/10.1007/s12525-022-00603-6.

Guleria, P., Naga Srinivasu, P., Ahmed, S., Almusallam, N., & Alarfaj, F. K. (2022). XAI framework for cardiovascular disease prediction using classification techniques. *Electronics,* 11(24), 4086. https://doi.org/10.3390/electronics11244086.

Han, H., & Liu, X. (2021). The challenges of explainable AI in biomedical data science. *BMC Bioinformatics,* 22(Suppl 12), 443. https://doi.org/10.1186/s12859-021-04368-1.

Holzinger, A., Saranti, A., Molnar, C., Biecek, P., &Samek, W. (2022). *Explainable AI Methods – A Brief Overview. In xxAI – Beyond Explainable AI. xxAI 2020.* Switzerland, Springer. https://doi.org/10.1007/978-3-031-04083-2_2.

Jiménez-Luna, J., Grisoni, F., & Schneider, G. (2020). Drug discovery with explainable artificial intelligence. *Nature Machine Intelligence*, 2, 573–584. https://doi.org/10.1038/s42 256-020-00236-4.

Joyce, D. W., Kormilitzin, A., Smith, K. A., Cipriani, A. (2023). Explainable artificial intelligence for mental health through transparency and interpretability for understandability. *npj Digital Medicine*, 6, 6. https://doi.org/10.1038/s41746-023-00751-9.

Keyl, P., Bischoff, P., Dernbach, G., Bockmayr, M., Fritz, R., Horst, D., Blüthgen, N., Montavon, G., Müller, K-R., & Klauschen, F. (2023). Single-cell gene regulatory network prediction by explainable AI. *Nucleic Acids Research*, 5(4), 1–14. https://doi.org/10.1093/nar/gkac1212.

Kim M-Y, Atakishiyev S, Babiker HKB, Farruque N, Goebel R, Zaïane OR, Motallebi M-H, Rabelo J, Syed T, Yao H, Chun P. A (2021). A Multi-component framework for the analysis and design of explainable artificial intelligence. *Machine Learning and Knowledge Extraction*, 3(4), 900–921. https://doi.org/10.3390/make3040045.

Kimber, T. B., Chen, Y., & Volkamer, A. (2021). Deep learning in virtual screening: recent applications and developments. *International Journal of Molecular Sciences*, 22(9), 4435. https://doi.org/10.3390/ijms22094435.

Krajna, A., Kovac, M., Brcic, M., & Šarčević, A. (2022). *Explainable Artificial Intelligence: An Updated Perspective*. In 2022 45th Jubilee International Convention on Information, Communication and Electronic Technology (MIPRO) (pp. 859–864). Opatija, Croatia. doi: 10.23919/MIPRO55190.2022.9803681.

Létinier, L., Jouganous, J., Benkebil, M., Bel-Létoile, A., Goehrs, C., Singier, A., … Pariente, A. (2021). Artificial intelligence for unstructured healthcare data: Application to coding of patient reporting of adverse drug reactions. *Clinical Pharmacology & Therapeutics*, 110(2), 392–400. https://doi.org/10.1002/cpt.2266.

Li, S., Wu, S., Wang, L., Li, F., Jiang, H., & Bai, F. (2022). Recent advances in predicting protein-protein interactions with the aid of artificial intelligence algorithms. *Current Opinion in Structural Biology*, 73, 102344. https://doi.org/10.1016/j.sbi.2022.102344.

Lin, R., & Wichadakul, D. (2022). Interpretable deep learning model reveals subsequences of various functions for long Non-coding RNA identification. *Frontiers in Genetics*, 13, 876721. https://doi.org/10.3389/fgene.2022.876721.

Liu, L., Meng, Q., Weng, C., Lu, Q., Wang, T., Wen Y. (2022). Explainable deep transfer learning model for disease risk prediction using high-dimensional genomic data. *PLOS Computational Biology*, 18(7), e1010328. https://doi.org/10.1371/journal.pcbi.1010328.

Lötsch, J., Kringel, D., & Ultsch, A. (2022). Explainable artificial intelligence (XAI) in biomedicine: Making AI decisions trustworthy for physicians and patients. *BioMedInformatics*, 2(1), 1–17. https://doi.org/10.3390/biomedinformatics2010001.

Lukyanenko, R., Castellanos, A., Storey, V. C., Castillo, A., Tremblay, M. C., & Parsons, J. (2020). Superimposition: augmenting machine learning outputs with conceptual models for explainable AI. In *Advances in Conceptual Modeling* (pp. 12584). ER 2020. Switzerland: Springer. https://doi.org/10.1007/978-3-030-65847-2_3.

Madan, S., Demina, V., Stapf, M., Ernst, O., Fröhlich, H. (2022). Accurate prediction of virus-host protein-protein interactions via a Siamese neural network using deep protein sequence embeddings. *Patterns*, 3(9), 1–11. https://doi.org/10.1016/j.patter.2022.100551.

Masuda, T., & Mimori, K. (2022). Artificial intelligence-assisted drug repurposing via "chemical-induced gene expression ranking". *Patterns (NY)*, 3(4), 100470. https://doi.org/10.1016/j.patter.2022.100470.

Mertes, S., Huber, T., Weitz, K., Heimerl, A., & André, E. (2022). GANterfactual-counterfactual explanations for medical non-experts using generative adversarial learning. *Frontiers in Artificial Intelligence*, 5, 825565. https://doi.org/10.3389/frai.2022.825565.

Montavon, G., Binder, A., Lapuschkin, S., Samek, W., & Müller, KR. (2019). Layer-Wise Relevance Propagation: An Overview. In *Explainable AI: Interpreting, Explaining and Visualizing Deep Learning* (pp. 11700). Switzerland: Springer. https://doi.org/10.1007/978-3-030-28954-6_10.

Muschalik, M., Fumagalli, F., Hammer, B., Hüllermeier, E. (2022). agnostic explanation of model change based on feature importance. *Künstliche Intelligenz*, 36, 211–224. https://doi.org/10.1007/s13218-022-00766-6.

Pakhrin, S. C., Shrestha, B., Adhikari, B., & Kc, D. B. (2021). deep learning-based advances in protein structure prediction. *International Journal of Molecular Sciences*, 22(11), 5553. https://doi.org/10.3390/ijms22115553

Ras, G., Xie, N., van Gerven, M., & Doran, D. (2022). Explainable deep learning: A field guide for the uninitiated. *Journal of Artificial Intelligence Research*, 73, 329–397. https://doi.org/10.1613/jair.1.13200

Rout, R. K., Umer, S., Sheikh, S., & Sangal, A. L. (Eds.). (2022). *Artificial Intelligence Technologies for Computational Biology* (1st ed.). USA: CRC Press. https://doi.org/10.1201/9781003246688.

Saeed, W. & Omlin. (2023). Explainable AI (XAI): A systematic meta-survey of current challenges and future opportunities. *Knowledge-Based Systems*, 263. https://doi.org/10.1016/j.knosys.2023.110273.

Sai, A. M. A., Sai Eswar, K. L., Sai Harshith, K. S., Raghavendra, P., Kiran, G. Y., & M. V. (2022). *Study of Lasso and Ridge Regression using ADMM*. In 2022 2nd International Conference on Intelligent Technologies (CONIT) (pp. 1–8), Hubli, India. doi: 10.1109/CONIT55038.2022.9847706

Sapoval, N., Aghazadeh, A., Nute, M. G., Antunes, D. A., Balaji, A., Baraniuk, R., Barberan, C. J., Dannenfelser, R., Dun, C., Edrisi, M., Elworth, R. A. L., Kille, B., Kyrillidis, A., Nakhleh, L., Wolfe, C. R., Yan, Z., Yao, V., & Treangen, T. J. (2022). Current progress and open challenges for applying deep learning across the biosciences. *Nature Communications*, 13, 1728.

Shi, P., Gangopadhyay, A., & Yu, P. (2022). *LIVE: A Local Interpretable model-agnostic Visualizations and Explanations*. In 2022 IEEE 10th International Conference on Healthcare Informatics (ICHI) (pp. 245–254), Rochester, MN, USA. doi: 10.1109/ICHI54592.2022.00045.

Thiagarajan, J. J., Thopalli, K., Rajan, D., Turaga, P. (2022). Training calibration-based counterfactual explainers for deep learning models in medical image analysis. *Science Report*, 12, 597. https://doi.org/10.1038/s41598-021-04529-5.

Ullah, I., Rios, A., Gala, V., & Mckeever, S. (2022). Explaining deep learning models for tabular data using layer-wise relevance propagation. *Applied Sciences*, 12(1), 136. https://doi.org/10.3390/app12010136.

Vo, T. H., Nguyen, N. T. K., Kha, Q. H., & Le, N. Q. K. (2022). On the road to explainable AI in drug-drug interactions prediction: A systematic review. *Computational and Structural Biotechnology Journal*, 20, 2112–2123.

Watson, D. S. (2022). Interpretable machine learning for genomics. *Human Genetics*, 141, 1499–1513. https://doi.org/10.1007/s00439-021-02387-9.

Yang, J., Li, Z., Wu, W. K. K., Yu, S., Xu, Z., Chu, Q., & Zhang, Q. (2022). Deep learning identifies explainable reasoning paths of mechanism of action for drug repurposing from multilayer biological network. *Briefings in Bioinformatics*, 23(6), bbac469. https://doi.org/10.1093/bib/bbac469.

Zacharias, J., von Zahn, M., Chen, J., Hinz, O. (2022). Designing a feature selection method based on explainable artificial intelligence. *Electron Markets*, 32, 2159–2184. https://doi.org/10.1007/s12525-022-00608-1.

Zhang, Y., Luo, M., Wu, P., Wu, S., Lee, T.-Y., & Bai, C. (2022). Application of computational biology and artificial intelligence in drug design. *International Journal of Molecular Sciences*, 23(21), 13568. https://doi.org/10.3390/ijms232113568.

Zhang, Z., Beck, M. W., Winkler, D. A., Huang, B., Sibanda, W., & Goyal, H. (2018). Opening the black box of neural networks: Methods for interpreting neural network models in clinical applications. *Annals of Translational Medicine*, 6(11), 216. doi: 10.21037/atm.2018.05.32.

Zhao, Y., Shao, J., & Asmann, Y. W. (2022). Assessment and optimization of explainable machine learning models applied to transcriptomic data. *Genomics, Proteomics & Bioinformatics*, 20(5), 899–911. https://doi.org/10.1016/j.gpb.2022.07.003.

Zhou, H., Wang, W., Jin, J., Zheng, Z., & Zhou, B. (2022). Graph neural network for protein-protein interaction prediction: A comparative study. Molecules, 27(18), 6135. https://doi.org/10.3390/molecules27186135.

Index

Printed in the United States
by Baker & Taylor Publisher Services